BIOTECHNOLOGY IN SOCIETY

Private Initiatives and Public Oversight

Pergamon Titles of Related Interest

Grunewald CHEMISTRY FOR THE FUTURE

Messel THE BIOLOGICAL MANIPULATION OF LIFE

Moo-Young, et al. ADVANCES IN BIOTECHNOLOGY:
Volume 1: Scientific and Engineering Principles
Volume 2: Fuels, Chemicals, Foods and Waste Treatment
Volume 3: Fermentation Products
Volume 4: Current Developments in Yeast Research

Moo-Young, et al. BIOMASS CONVERSION TECHNOLOGY

Moo-Young, et al. COMPREHENSIVE BIOTECHNOLOGY: PRINCIPLES,
METHODS AND APPLICATIONS

Moo-Young, et al. WASTE TREATMENT AND UTILIZATION

Morgan & Whelan RECOMBINANT DNA AND GENETIC
EXPERIMENTATION

OTA COMMERCIAL BIOTECHNOLOGY

Rothman, et al. BIOTECHNOLOGY

Zadrazil & Sponar DNA: RECOMBINATION, INTERACTIONS AND
REPAIR

Related Journals*

BIOTECHNOLOGY ADVANCES
CELLULAR AND MOLECULAR BIOLOGY
CHEMOSPHERE
CURRENT ADVANCES IN GENETICS AND MOLECULAR BIOLOGY
ENERGY CONVERSION AND MANAGEMENT
JOURNAL OF PHARMACEUTICAL & BIOMEDICAL ANALYSIS
MATHEMATICAL MODELLING
MOLECULAR ASPECTS OF MEDICINE
NUCLEAR AND CHEMICAL WASTE MANAGEMENT
PROGRESS IN FOOD AND NUTRITION SCIENCE
TECHNOLOGY IN SOCIETY

* Sample copies available on request

Sent With Our Compliments . . .

Pergamon Press, Inc.
Maxwell House, Fairview Park
Elmsford, New York 10523

BIOTECHNOLOGY IN SOCIETY
Private Initiatives and Public Oversight

Joseph G. Perpich
Meloy Laboratories
Guest Editor

Published in part as special issues of
Technology in Society, *An International Journal*

George Bugliarello
A. George Schillinger
Polytechnic Institute of New York
Editors

Martha Miller Willett
Polytechnic Institute of New York
Managing Editor

Pergamon Press
New York Oxford Toronto Sydney Frankfurt

Pergamon Press Offices:

U.S.A. Pergamon Press Inc., Maxwell House, Fairview Park,
Elmsford, New York 10523, U.S.A.

U.K. Pergamon Press Ltd., Headington Hill Hall,
Oxford OX3 0BW, England

CANADA Pergamon Press Canada Ltd., Suite 104, 150 Consumers Road,
Willowdale, Ontario M2J 1P9, Canada

AUSTRALIA Pergamon Press (Aust.) Pty. Ltd., P.O. Box 544,
Potts Point, NSW 2011, Australia

**FEDERAL REPUBLIC
OF GERMANY** Pergamon Press GmbH, Hammerweg 6,
D-6242 Kronberg-Taunus, Federal Republic of Germany

BRAZIL Pergamon Editora Ltda., Rua Eça de Queiros, 346,
CEP 04011, São Paulo, Brazil

JAPAN Pergamon Press Ltd., 8th Floor, Matsuoka Central Building,
1-7-1 Nishishinjuku, Shinjuku, Tokyo 160, Japan

**PEOPLE'S REPUBLIC
OF CHINA** Pergamon Press, Qianmen Hotel, Beijing,
People's Republic of China

First printing 1986

Library of Congress Cataloging in Publication Data
Main entry under title:

Biotechnology in society.

    "Published in part as special issues of Technology in
Society, an international journal, George Bugliarello,
A. George Schilliger . . . editors."
    1. Biotechnology--Social aspects. I. Perpich,
Joseph G. II. Technology in society.
TP248.2.B557 1985   303.4'83   85-21630
ISBN 0-08-033168-8

Printed in Great Britain by A. Wheaton & Co. Ltd., Exeter

Contents

Part VI

Foreword

Few people are as familiar with and have contributed as much to the development and growth of the new biotechnology industry as Joseph G. Perpich, the distinguished guest editor for the *Technology in Society* Special Series. A psychiatrist and attorney, Dr. Perpich is a graduate of the University of Minnesota Medical School and completed his internship at the university's hospitals, followed by a residency in psychiatry at the Massachusetts General Hospital and the National Institute of Mental Health. Under a mental health career development award from the NIMH, Dr. Perpich combined his legal and psychiatric training for one year as a Congressional Fellow on the Subcommittee on Health of the Senate Committee on Labor and Public Welfare, and the other as a law clerk to David Bazelon, then Chief Judge for the US Court of Appeals for the District of Columbia Circuit. He received his law degree in 1974 from the Georgetown University Law Center.

From 1974 to 1976, Dr. Perpich was responsible for the development of a program in legal and medical ethics at the Institute of Medicine, National Academy of Sciences. As Associate Director for Program Planning and Evaluation at the National Institutes of Health under NIH Director Donald S. Fredrickson from 1976 to 1981, he directed the staff effort on developing the NIH's Recombinant DNA Research Guidelines and related federal legislative, judicial and executive policies in this emerging field. His final assignment at the NIH was in the area of bringing industry further into the government-university partnership in health research, particularly in the area of biotechnology.

Until recently Vice President for Corporate Planning and Government Affairs for Genex Corporation, in January of 1984 Dr. Perpich joined Meloy Laboratories as Vice President for Planning and Development. A wholly owned subsidiary of Revlon, Meloy is one of several companies which make up Revlon Health Care. Meloy scientists conduct research, provide research support services, produce biological products for government agencies and private industry, and develop and produce diagnostic assays and reagents for use by hospitals and medical centers worldwide.

The editors of *Technology in Society* wish to express their deep appreciation to the many distinguished experts who contributed to this comprehensive series on biotechnology, and to extend their sincere gratitude to Dr. Perpich, whose experience, expertise and dedication made possible this valuable contribution to the biotechnology literature. Dr. Perpich has contributed a work that is an indispensable resource for persons interested in the "new biology" or who work in the industry that has been created by these revolutionary new discoveries.

George Bugliarello
A. George Schillinger

Preface and Acknowledgements

This book is an outgrowth of a series of articles that appeared in the journal, *Technology in Society*, from 1982 through 1984 with a focus on the wide ranging potential that the "new biology," as represented by biotechnology, holds for society. Biotechnology is comprised of many processes—fermentation and the emerging genetic engineering techniques, primarily cell fusion and recombinant DNA technology. In Part I, the opening article in this book, which did not originally appear in the *Technology in Society* series, presents an excellent introduction to the developing science, technology and commercial applications of industrial biotechnology.

The subsequent sections in this volume document and examine biotechnology's present and future relationships with societal institutions. In Part II, the authors examine the "new biology" from the perspective of research relationships and goals: among governmental groups, universities and industry. Part III reviews biotechnology's impact on policy formulation and implementation in the congressional, judicial and executive branches. Part IV examines the specific problems biotechnology poses to corporate, regulatory and patent law. Part V focuses on difficult and long-range ethical questions raised by this technology, particularly as these techniques are applied to modify and construct new genes in humans. In addition, in this chapter, there is an examination of the international impact of biotechnology and an evaluation of public and press perceptions as they relate to the consequences of biotechnology on societal institutions.

The concluding section, (Part VI) which did not originally appear in the *Technology in Society* series, deals with a major policy area of regulatory oversight, namely, federal export regulations governing biotechnology processes and products. The authors' presentations focus on the two factors driving the move for export regulations—the possibility of biotechnological warfare and the prospect of well-coordinated foreign competition.

The authors in this volume address the challenges posed by biotechnology as well as the enormous promise to meet what Dr. Krause, in his article in this volume, calls the "trinity of despair"—hunger, disease and insufficient resources to meet an expanding world population. All involved in biotechnology recognize the need for the open atmosphere of inquiry that allows, in Dr. Lewis Thomas' words, a "cascade of surprises" to occur. And all recognize the necessity of support for basic research from all major partners: government, academia and industry, and, concomitantly, the indispensable need for circumspection and public oversight in the applications of this technology to humans and the environment and the potential misuse for biological warfare purposes.

I would like to thank George Bugliarello and A. George Schillinger, the editors of the Journal, who invited me to serve as the guest editor and strongly supported and encouraged the development of the series in the Journal. I would also like to thank each of the contributors for their participation, invaluable assistance and dedication to the development and publication of this series of articles. I also wish to acknowledge the advice, cheerful support and patience of the Journal's managing editor, Martha Willett, and Pergamon's editor, Elyse Dubin. Especially, I thank Shellie Roth, Director of Corporate Communications, Genex Corporation, for her excellent editorial support for the series; my former secretary, Melissa Andrews, who worked unfailingly to maintain an orderly process for my work with the contributors and my editors. With the publication of the *Technology in Society* series in this volume, I wish to thank Thomas Anthony, Publishing Manager at Pergamon Press, Inc., who worked closely with me in the development and publication of this volume and my present secretary, Evalyn Sheehan, who gave generously of her time and talents to assist me in preparing these articles for publication.

Finally, I dedicate this volume to David L. Bazelon, Senior Circuit Judge of the U.S. Court of Appeals for the District of Columbia (retired), who taught me the importance of public participation in government policy making and the development of a public record, and to Donald S. Fredrickson, President of the Howard Hughes Medical Institute and former Director of the National Institutes of Health, who taught me that the "two cultures" of science and society can be joined when the scientists and their institutions are led by individuals committed to ensuring that the development of technology proceeds hand-in-hand with full public accountability in order to preserve the public good-will on which all of science depends.

Part I

1
Introduction
Joseph G. Perpich

The articles which appear in this volume have as a major theme the roles and responsibilities of government, universities, and industry in the development and oversight of biotechnology, not only within the United States, but internationally as well. In the opening article, based in part on a speech given at the Commonwealth Club in San Francisco in 1984, Dr. Ronald Cape addresses the respective roles of government, university, and industry in biotechnology R&D and product development, and the government's critical role in maintaining US leadership in biotechnology.

Ronald Cape

Ronald Cape is Chairman and Chief Executive Officer of Cetus Corporation. He holds an A.B. in Chemistry from Princeton, an M.B.A. from Harvard University's Graduate School of Business Administration, and a Ph.D. in biochemistry from McGill University. His post-doctoral work was conducted at the molecular biology and virus laboratory at the University of California, Berkeley. In 1971, he co-founded the Cetus Corporation. Under his direction, Cetus established itself as a pioneer in the emerging field of biotechnology, with a particular focus on health care and agricultural products and processes. Dr. Cape was an active spokesman for the industry when the NIH was drafting its guidelines in 1976–1977, frequently testifying on behalf of the new biotechnology firms and calling national attention to this growing private sector. As the President of the Industrial Biotechnology Association (1983–1985), he continued to serve as an advocate for industrial organizations with active programs in biotechnology.

Dr. Cape describes biotechnology's industrial revolution, which, in large part, is fueled by discoveries in biomedical research funded by the National Institutes of Health. Although the United States pioneered in the industrialization of genetic engineering, Dr. Cape cautions that the United States may lose its national leadership if federal support for basic research does not keep pace with the enormous R&D potential that exists today among government, the universities, and the biotechnology industry. In his view, federal research support is lagging behind scientific opportunities. Accordingly, universities are aggressively seeking industrial funds and promoting technology transfer, which, in turn, has created an identity crisis for the academic community. (Dr. Donald Kennedy's paper addresses this issue in the next series of articles in this volume.) The newer biotechnology firms, which developed from an enormous infusion of private capital over the past decade, face narrower opportunities for product development when funding for basic research at universities is diminished. Dr. Cape asserts that the foundations of this industry rest on university-based research and graduate education in biotechnology disciplines.

Dr. Cape also calls for the government to maintain its traditional funding role in light of Japan's long-term commitment to industrial biotechnology. (The last series of articles in this volume is devoted to the issues surrounding international trade and foreign economic and national security competition.) He outlines the Japanese bid for dominance in biotechnology's lucrative future, supported by that country's government, both at the basic research level, and again at the point where risk-taking entrepreneurs in the American model usually invest money. Dr. Cape calls for restoration of the partnership among government, academic, and private sectors that has been so successful in the past. He concludes by urging a coalition be formed among action-oriented businessmen and academics to garner the public support necessary for an enhanced long-term federal commitment to fund basic research and education in the biotechnology disciplines at US academic institutions.

Future Prospects in Biotechnology:
A Challenge to United States Leadership

Ronald E. Cape

Biotechnology is a business with a short past—begat from a scientific genealogy not much more than two decades old. It is an industry with a robust present, attracting investment money and generating tangible products. Most importantly, it is a national opportunity, slipping away into an uncertain future. In a figurative sense, biotechnology is a miracle that uses nature's most elegant solutions to solve mankind's most puzzling mysteries. In a literal sense, it is a set piece of how government, academia, and private enterprise in America have traditionally cooperated in an unbeatable formula for leadership in research and development. Unfortunately, biotechnology is also fast becoming the tale of how political myopia can dismantle that successful formula.

To the lay public, biotechnology may still be a nebulous, mysterious science, mysterious even though it has been featured in everything from small-town newspapers to *Scientific American* to *Pravda*. Popular opinion equates it to science fiction-like futuristic scenarios—promising but not delivering. In fact, industry-watchers have often dismissed the business of biotechnology as altogether impossible. They have specialized in coming up with barriers and obstacles to explain why biotechnology cannot, and will not, do what those of us in development say it can do. But unlike some other advanced technology businesses, biotechnology is not only achieving its goals, in many cases it is achieving them earlier than expected. Cetus Corporation, for example, went from isolating the gene, the chemical instruction controlling heredity known as DNA, for a possible new anti-cancer and anti-AIDS therapeutic to testing the pure drug in human beings in only eleven months. The drug is interleukin-2, and it is an amazing biotechnological accomplishment to move from a gleam in a scientist's eye or a drop of liquid in a test tube to clinical trials in less than one year. But that is the power of biotechnology; its products are already delivering, more are coming, and they are literally going to change the way people live.

Reduced to its briefest description, the power of biotechnology emerges from two things. First, it depends on knowledge about the nature of life itself and, second, it is able to circumvent basic differences among organisms. Understanding it necessitates explaining a little about the inborn codes that make some folks tall, form an elephant's trunk, or ensure that insects have six legs: the science of genetics.

All life is organized into groups, or species. Although it may seem just a convenience for biologists, the arrangement stems from the fact that genes (the factors of heredity) mix successfully only when males and females of the same species mate.

Up to now, Mother Nature has dictated that you cannot breed across species lines. Thus no one expects to see elephants romancing mosquitoes or white rabbits courting green frogs — at least not productively.

However some natural gene crossing does occur. Most people know that when a donkey representing one species breeds with a horse representing another, a mule results. Biotechnology finesses the exception into a commonplace event, so that genes literally jump between species of mice or frogs. That is part of what gets people so excited — and concerned — about genetic engineering. Biotechnologists are simply getting Mother Nature to move faster and in new ways, albeit in a controlled, precise fashion that is both design and engineering — or genetic engineering.

The engineering takes place by dispensing with the barrier (in this case, at least) of sexual mating, and instead going straight to the chemical codes of heredity — DNA. DNA is the instruction, the blueprint our cells use to do and make things. Those who see the world with hazel eyes, for example, do so because their DNA issued chemical instructions for that color. The same goes for curly hair, white blood cells, digestive enzymes, or fat ankles. Because DNA is just a genetic language, common to all living creatures, it is possible to take fragments of one life form's genetic information and splice it into the genetic information of another. Using molecular biology techniques, scientists recombine the DNA pieces they isolate for their special properties, hence the full name: "recombinant DNA engineering." Usually, the donor DNA is inserted into a host cell, where it reproduces and works as it was meant to, no matter what the organism. Again, DNA is a universal genetic language. And in contrast to the random sexual transfer of genes, laboratory "gene moving" is quite precise.

All of this is more than just a neat trick. Biotechnologists do not combine human DNA with, say, a lowly yeast cell for the sheer novelty of it. For one thing, they do it to produce extremely rare human substances — proteins like interferon, growth hormone, or interleukin-2 — in very pure form but in great quantities and at reasonable cost. All human bodies contain the genetic instructions to make these same substances naturally, but all the interferon that all of humanity could produce, for instance, only adds up to a pinch.

Instead, biotechnology allows industry to splice the interferon DNA into some cooperative little bacterium or yeast cell that reproduces quickly and abundantly. The biotechnological process of yeast fermentation in brewing beer or making wine works on the same principle. However interferon is a bit more precious than a six pack (or even a good zinfandel); five years ago before biotechnology's help, it figured out to $22 billion a pound. Now it is in human clinical trials in the fight against cancer because we have large enough quantities of this substance to experiment with what the body naturally produces to defend itself. Mind you, it will be some combination of luck and miracle if we find a cancer cure with these first product efforts. But that is no reason for discouragement, because it is just a matter of time. Genetically engineered human insulin, for instance, is already on the market, and the industry is also at work on vaccines against hepatitis and herpes. Of course, the real pay-off will come when effective products for treating cancer and heart disease emerge from the laboratories. The fact is, genetic engineering's approach makes sense for any number of beneficial applications.

Genetic engineering's approach also makes good business sense, which means that biotechnology is positioned to make a lot of money. The Commerce Department has predicted that, though the market for genetically engineered products is less than $100 million today, by the 1990s that market will increase to tens of billions as a result of its expected breakthroughs. It will, that is, only if the key players of government, academia and industry can reverse a trend that is dismantling an American tradition — the mutually beneficial exchange that made biotechnology possible in the first place.

The present American leadership in biotechnology as private enterprise has its roots in a decades-old American leadership role in basic research. Heavy investment in basic research, sooner or later, pays off handsomely. Between 1945 and 1975, the United States spent billions on basic research in universities, research centers, and especially at the federal government's National Institutes of Health. As a direct result of that investment, scientists in this country slowly, carefully inched forward in what they understood about life. Their work was far from the marketplace; those scientists were not intent on developing commercial products. That was simply not their motivation. Instead, like research scientists now, they were inspired, excited people who enjoyed their work immensely. They spent their lives searching for fundamental truths. And they were also human and very competitive about international recognition for their work.

Yet the progress in biotechnology product development today would be impossible without the crucial breakthroughs in fundamental research that took place since 1945, long before anybody ever dreamed of the practical applications. Authorities throughout the world share this opinion. Consider, for example, the evidence in terms of Nobel Prizes awarded in this area. US scientists involved in basic biotechnology research have literally won dozens. Consider, too, an indication that American leadership is faltering. Today the greatest challenge to biotechnology in the United States comes from Japan, although that country has earned no Nobel Prizes for basic research in biotechnology. What makes that challenge possible then? Instead of continuing to pioneer the industrialization of genetic engineering, American leadership may fade because federal funding for basic research is not increasing.

On the contrary, instead of doubling or tripling, it is decreasing. A *New York Times* article reported that the Reagan administration had decided to reduce federal support for biomedical research in FY 1985 below the levels intended by Congress. The move would have cut the number of NIH competitive grants awarded in 1985 by 23 percent. In response to questions, one Administration official said, "Some scientists are fussing. But everybody who loses a subsidy fusses." Such an attitude is clearly inconsistent with the repeated statements of Dr. George A. Keyworth II, the president's science advisor, that federal basic research support is an investment in the nation's future security, economy, and quality of life. Only a legitimate investment, and not a "subsidy," pays the kind of dividends we are collecting today.

What I describe is more than a malaise; it is a crisis that demands action. America needs to restore its national commitment to support basic biological research. Delving a little more into the different interests and problems of the various communities involved will lead to some preliminary ideas on what to do.

First, the universities have profound financial, psychological, organizational, and administrative problems of both a short- and long-term nature. Industry, on the other hand, currently has few short-term problems, but will falter in the long run if the other communities involved do not solve their own problems. And, long-term problems will confront politicians if they do not address the short-term problems immediately. Finally, the United States will suffer all around if the players do not soon agree on a remedy. All the communities are mutually interdependent; if one falls, the others will suffer as well.

The universities contribute teaching, basic research, and training. Industry provides jobs and opportunities to apply basic research to the production of goods and services, which results in profit. Taxes on those profits ideally return to the universities in the form of federal support. In addition, the patent system rewards invention in both academia and industry. And that is the American research and development system in a nutshell.

Today, the universities are feeling the squeeze of budgetary belt-tightening and feeling it badly. Individual investigators especially experience increasing difficulty in getting research grants approved. As a result, universities are turning to licensing and patent activity, as well as currying new relationships with industry. Not that financial problems are the only ones plaguing the academic community. There is a crisis of institutional identity that results from both the decline in federal support and the increased attention from industry. Historically, resources have changed hands among universities, professors, and companies without significant problems. However the identity crisis arises from the birth of the so-called DNA companies with their enormous sums of money, together with the equally large financial commitments to biotechnology made by major multinational companies and even some foreign governments.

Accordingly, universities are openly courting large companies and, to put it bluntly, selling the services of entire departments to them. Professors, usually the best ones, are diverting more of their intellectual, creative energy away from university research and teaching. And if biotechnology is already yielding a rich harvest of benefits, then we are consuming our own seed corn. When the commitment of a professor to a company becomes overwhelming, it is better to resolve the ambiguity by leaving academia and declaring oneself a full-time employee of the company. Two conspicuous examples of this straightforward solution are Professor Walter Gilbert's leaving Harvard for Biogen, and Professor Winston Brill's leaving the University of Wisconsin for the Cetus-Madison agricultural operation. However, the point is, a previously healthy equilibrium that brought the United States to world technological leadership has been seriously disturbed.

For the time being, industry might be able to ignore the straitened times universities face. Because of the momentum that still exists from early research, the private sector is working in a textbook paradigm: A healthy stream of capital has been flowing into genetic engineering over the last ten years. Ask why and the answer is simple: Everyone agrees—the public sector, foreign companies, foreign governments —with industry leaders that by the 1990s the market for genetically engineered products will be staggeringly lucrative. But while biotechnology is no longer a high-risk business, it remains a "high-patience" one. It requires a long-term commitment, and everyone from entrepreneurial companies to multinational corporations has acted on it—and without government encouragement or help it should be added.

But the need to maintain the academic resource that nurtures future product development still remains.

The government, on the other hand, has demonstrated its interest in biotechnology in disappointing ways, either with decreased funding or increased regulation. Nonetheless, without the traditional federal support to research, biotechnology in this country could slowly settle into a solid second place behind Japan. Japan's biotechnology activity already exceeds 7 percent of its gross national product, the highest percentage of any nation on earth. Moreover, Japan's commitment to biotechnology is a long-term one, and an excellent match for the science. Given the dismal state of the US commitment on the other hand, we can dread the day when biotechnology business will be an American saga of the also-ran. Because it is business and not research that turns out commercial quantities of products like interferon, we lose out on the dominant technology of the next twenty years — a technology that can save lives, preserve the environment, feed the hungry through agricultural applications, and revolutionize a host of other industries from detergents to mining. Finally, along with world technological leadership, we sacrifice the obviously significant economic advantages.

So what, however tentatively, can be done to convince Washington to restore and increase substantial support for basic research in biology? That is a tough task to tackle, for getting things done in the political arena is a science all by itself, and one in which most conventional scientists do not excel. Nevertheless, I submit that a coalition, formed among action-oriented businessmen and articulate academics, together with a small number of committed government people from both Congress and the executive branch could come up with a plan and carry it out. This is not idle dreaming. Consider the skepticism that would have resulted if anyone had told of a plan in the middle of the Carter administration to reduce the capital gains tax from almost 50 percent to a figure in the 20 percent range. But a small group who knew how to articulate a position and garner support for it accomplished that unlikely objective. Keep in mind that I am not representing some narrow, special interest here. I am not asking money for my company, or even for the biotechnology industry. I am asking for money for basic research at our country's universities and at the National Institutes of Health, and at other centers of basic research.

A more concrete proposal is to examine the biotechnology institutes established by the governments of West Germany, Japan, and the United Kingdom. These institutes are government-supported and bring together academic and industry investigators for cross-disciplinary research activities. Such institutes could be established by the federal government in several of the relevant research agencies (such as the National Institutes of Health, the National Science Foundation, and the Department of Agriculture) to promote biotechnology research and development in health, chemicals, and agriculture, respectively, among academic and industrial institutions.

The rewards for the commercialization of biotechnology are profound. The United States will only have itself to blame if it does not try to maintain its leadership in research and development by keeping the biology research base primed to meet the enormous opportunities that lie ahead. Basic research is the wellspring from which everything else flows, and with an all-out effort we in the private sector can do what we do best — take the research and run with it, to benefit all humankind.

Part II

3
Introduction
Joseph G. Perpich

The articles in the next several chapters in this volume focus on the wide-ranging potential the "new biology" holds for society. Although most of the authors' discussions will focus on recombinant DNA technology, it is important to be cognizant that broader biological techniques are involved, such as the use of immobilized enzymes in continuous flow bioreactors. To the reader uninitiated in such terminology, the articles appearing in this section will be a good introduction to the developing science, technology, and commercial applications as well as to the societal issues surrounding them.

The nascent biotechnology industry combines the skills of academic research scientists, fruits of federally funded basic research, and the venture-capitalist's dollars. But this unique blend is now challenging traditional government–university–industry relationships, raising questions among the partners that, in their resolution, are likely to effect basic changes in those research relationships. Therefore, we begin by examining biotechnology as a whole and its particular use of recombinant DNA technology. This is an important starting point, for the research and development programs now present in universities, government agencies, and the industry itself are beginning to reap commercial applications that will touch many aspects of our lives: health care, agriculture, food-processing, and environmental management—to name but a few.

Donald Kennedy

The first contributor is Dr. Donald Kennedy, President of Stanford University, who received his undergraduate, master's, and Ph.D. degrees from Harvard University. His academic career has been largely at Stanford University, where he was Professor and Chairman of the Department of Biological Sciences. From 1977 to 1979, Dr. Kennedy served as Commissioner of the Food and Drug Administration. In 1979 he returned to Stanford University as Vice President and Provost, becoming President of that institution in 1980.

This blend of scientific, academic, and government experience puts Dr. Kennedy in the forefront of the debate now in progress over emerging changes in research collaboration among government, academia, and industry. Indeed, Stanford University is one of the nation's leading academic sites for research in recombinant DNA technology. Several Nobel laureates, including Paul Berg, who contributed to the fundamental understanding of recombinant DNA research, reside there.

Stanford has also been in the forefront of the examination of current industry–university collaborative research agreements. Since assuming the Presidential post at Stanford, Dr. Kennedy has taken a leadership role toward examining the fundamental nature of existing and future government–university–industry collaborative research. He has looked carefully at the need to enhance those research relationships while respecting the particular roles each of the partners must play to fulfill its mission. In fact, Dr. Kennedy was instrumental in organizing a three-day meeting at Pajaro Dunes in 1982, bringing together the presidents of Harvard, the Massachusetts Institute of Technology, the University of California, the California Institute of Technology, and Stanford to discuss the issue. That convocation of 35 faculty scientists, industrial representatives, and university administrators helped to set the current agenda for debate on the commercialization of the new biology.

A draft statement released after the Pajaro Dunes meeting noted: "Research of the past several decades, through enlightened public support, has profoundly advanced the understanding of life processes. A new biotechnology of extraordinary promise has emerged. . . . The chain of progress from basic research to useful applications necessarily involves universities and industry. For the promise to be fulfilled, all links in the chain must be strong."

Dr. Kennedy's article for this book, which is based on the Wells Fellowship lecture he gave at Yale University in 1982, offers principles and policies to guide universities and their faculties as they consider these overriding issues. After reviewing the nature of Federal support for biological research over the past 40 years and the attendant roles of government, academia and industry in the innovation process, Dr. Kennedy notes the fundamental rearrangement in the social sponsorship of discovery that is now taking place. He addresses what he sees as a "revolutionary compression" of the innovation process, and its concomitant challenges to the governance of the university and its faculty.

Richard M. Krause

Richard M. Krause is Dean of the Emory University School of Medicine and Woodruff Professor of Medicine. He was former Director of the National Institute of Allergy and Infectious Diseases (NIAID) at the National Institutes of Health (NIH). A physician, immunologist and microbiologist, Dr. Krause received his training at the Case Western Reserve University Medical School. Following faculty appointments at the Washington University School of Medicine in St. Louis and the Rockefeller University, he served as Director of the NIAID from 1975 to 1984. NIAID supports much of NIH's fundamental research in recombinant DNA technology, cell fusion and related areas. Located within the NIAID is the Office of Recombinant DNA Activities, with responsibility for the overall review and implementation of the NIH Guidelines for Recombinant DNA Research.

Dr. Krause is the author of *The Restless Tide: The Persistent Challenge of the Microbial World*, a series of essays reviewing the remarkable advances of medicine and science from the time of Pasteur.

Based on a lecture he presented at a 1982 seminar on "Science and Our Society" at the University of Georgia, Dr. Krause's article here expands on this essay. He describes the revolutionary fever of the last 40 years of biological discovery, and asks how the resulting cascade of information can deal with what he calls the "trinity of despair": hunger, disease, and insufficient resources to support an expanding population. The dimensions of these problems are worldwide, he notes. Citing T.S. Eliot's statement "the whole world is our hospital," Dr. Krause outlines a challenging strategy for the United States to team up with other democracies to match the promise of the new biology with the major tasks before society.

J. Leslie Glick

J. Leslie Glick, Chairman of the Board and Chief Executive Officer of Genex Corporation, brings an industrial perspective to the issues raised by Drs. Kennedy and Krause. Dr. Glick received his undergraduate and graduate degrees from Columbia University. After serving as chairman of the physiology department at the Roswell Park Division of the State University of New York at Buffalo, he entered industry, eventually becoming President and Chairman of the Board of Associated Biomedic Systems, a biologicals manufacturer and research and development company. Dr. Glick has been Chief Executive Officer of Genex, a biotechnology firm, since its inception in 1977; he also served as the first President of the Industrial Biotechnology Association (1981–1983), created in 1981 for industrial organizations active in biotechnology programs.

In his article, Dr. Glick addresses the broad impact biotechnology will have on many segments of society's traditional industries. Beginning with a description of conventional fermentation technology, the article details how our older technologies blended with new findings in molecular biology to form what is now termed the "biotechnology industry." Dr. Glick discusses both near-term and long-range prospects for genetic engineering in industry and includes economic assessments and models for his scientific prospects. He notes that one of biotechnology's major accomplishments will be the custom design of proteins and enzymes, a target that, when reached, will indicate that the new biology will have outgrown its infancy.

Thus, contributors to this section of articles, which originally appeared in *Technology in Society* in 1982 (Vol. 4, No. 4), name the issues and themes that will be prominent in subsequent chapters of the book. For it is from these institutions and research relationships that biotechnology will effect its mark on society, having been elaborated and challenged by governmental, private sector, and academic institutions.

4

The Social Sponsorship of Innovation

Donald Kennedy

US society is one that was built on innovation. The Constitution recognizes that; it specifically grants to the Congress the power "to promote the Progress of Science and useful Arts, by securing for limited time, to Authors and Inventors the exclusive Right to their respective Writings and Discoveries." Accordingly, the US has, throughout recent history, incorporated into its social institutions arrangements for the sponsorship of innovation. Of prime importance was the mid-19th century development, through the Morrill Act and the Hatch Act, of the land-grant university system and the network of State Agricultural Experiment Stations. That created an automatic flow of public funds for agricultural research, and a group of institutions responsible for its diffusion and application. The base for science of a more fundamental kind was laid with the grafting of the German graduate university concept onto the tradition of liberal humanistic education that had been a British inheritance. Finally—still before the turn of the twentieth century—some relatively new institutions of private character were declaring a strong commitment to the practical arts, including engineering. Among these were first Cornell and MIT, and later Stanford.

The institutional apparatus for doing science was, except for agriculture, a vessel relatively empty of public funds until World War II. The need to do something with it had been recognized—if not widely, at least by an astute few. Shortly after building the house in which I now live, Herbert Hoover, an engineer-scientist of considerable distinction, was named Secretary of Commerce. His major initiative

This paper is adapted from the Wells Lecture in Jonathan Edwards College of Yale University in 1982. The author gratefully acknowledges the support of the Wells Fellowship.

was an effort to raise a fund of ten million dollars from American industry to support basic research in the nation's universities. Hoover argued tirelessly for his project, on the grounds that American industry needed the fundamental research as well as the trained scientists that could be produced by academic institutions. He told industry leaders that they would be losing a form of intellectual capital if they did not make it possible for able researchers in the universities to be relieved of some of their teaching obligations, and equipped to do first-rate scientific work. It was the first effort by a national leader, as far as I know, to link academic research explicitly to productivity objectives.

Hoover's campaign fell far short of its goal, although it did help to create the initial support for the National Research Council and a program that helped keep academic science alive during the Depression. But following World War II, there was a dramatic realization of the Hoover vision — through public instead of private funds. The extraordinary scientific apparatus built for military purposes was transformed, under the influence of Vannevar Bush and others, into a system for the support of peacetime science in the universities. That support began largely with the Department of Defense, especially the Office of Naval Research; but in the early 1950s it was mostly transferred to the National Science Foundation and the rapidly growing programs of the National Institutes of Health. As these organizations took over the sponsorship of fundamental research, Defense and other mission-oriented agencies concentrated on programs related to their special areas. The outcome was a complicated patchwork of affiliation. In a large midwestern state university, for example, one would find a department of engineering supported largely by Defense agencies, a physics department mainly dependent on NSF and AEC monies, and two departments of biochemistry — one formula-funded by Agriculture and the other almost entirely dependent upon NIH biomedical research money.

The Trajectory of Innovation

Despite these complexities, the public support of fundamental science has become an effective instrument for meeting national goals, and it has been successfully integrated with the country's commercial style of product development. As a result of more than three decades of that coexistence, there has evolved a fairly standard historical sequence of innovation, within which particular kinds of institutional sponsors play defined roles. The innovative process normally begins in a university or government laboratory. This first stage is characterized by loose, informal organization and very open communication, including quick publication of all the details of an experiment. Stage I is publicly funded and oriented toward the discovery and explanation of *phenomena*. Examples of institutions in which it might take place would be university departments of biology or chemistry or physics, a laboratory in an NIH Institute, or a special industrial organization like Bell Telephone Laboratories.

The second stage is best called *application*. It is focused upon *processes*, and takes place in various settings: applied institutes, some university departments (of engineering, for example), nonprofits (like SRI International or Battelle), and industrial laboratories. There is a mix of public and private funding, and environments that are variable with respect to proprietary secrecy.

In the third stage, *development,* attention is given to practical application — including such matters as scale, rates, and means of economical production. The innovative emphasis is on *products;* funding is by private risk capital, and the environment tends to be closed for proprietary reasons and tightly managed. Essentially all such work takes place in commercial laboratories.

Individual innovations have histories; though these are seldom simple, they usually follow a trajectory that takes them along a sequence from phenomenon through process to product. That trajectory is concave upwards, because the rate of development normally quickens as the application is first glimpsed and then seen as certain. The ordinate can be expressed in terms of proportion of private financing, or degree of secrecy, or tightness of management control. Although the three stages just described are arbitrary, they do signify sharp differences in style, control, and financing. And the contrast between the two ends of the curve is dramatic: Research of the most fundamental sort is an entirely different social process from product innovation.

Public Understanding of the Innovation Process

That point may not seem worth all the attention I have been giving it. But in fact it has been the subject of important public misunderstanding and, as a consequence, of serious policy errors. A recent case in point is the so-called Small Business Innovation Act passed by Congress in July of 1982. This legislation sets aside 1¼ % of the research budget of each major federal research agency and establishes Small Business Innovation programs that will channel these reserves to qualifying small businesses. Like most statutory 'fixes,' this one was well-intentioned: It sought to increase the incentive to innovate in a sector of our economy that has been responsible for many of its best ideas, not to mention most of its new jobs. But, because it mistakes one kind of research for another, it will actually reduce total scientific productivity. Note that public funds will be removed from fundamental research and used to support product innovation.[1] The flow of new ideas *into* the stage at which application becomes possible will thereby be reduced. Furthermore, these public funds will be applied in an area in which, thanks to recent changes in the tax laws, abundant sources of private venture capital already exist to support good ideas. So the net effect will be a transfer of increasingly scarce public funds from basic science that can be supported in no other way to applied science that is already very attractive to private investment.

How, you might ask, could such a thing happen? It happened, I think, because of authentic confusion about how innovation happens and about what kinds of social arrangements have already been made for encouraging it. Small business is, of course, a good thing to be these days; indeed the Senate version of the Bill came through Committee in record time and was passed on the floor by a 90–0 vote before anyone in the scientific community woke up and looked seriously at the Bill.[2] As the *Washington Post* later editorialized, when a piece of legislation is passed in that way you *know* somebody isn't looking very hard. But the House gave the matter less perfunctory treatment, with some encouragement from us, and so proponents had to muster more serious arguments.

One sample of their logic is particularly revealing: the Small Business Committee prepared a list of innovations that could be credited to small businesses. Included on it was the invention of penicillin. I do not think the proponents deliberately ignored Alexander Fleming; my own experience in government always inclines me to favor confusion over conspiracy as an explanatory hypothesis. My point is that misrepresentations that would have produced an outburst of laughter among scientists passed almost without notice in the Congressional debate over this bill. To their credit, the Office of Management and Budget and the President's Science Adviser recognized what was happening and testified forcefully against the bill in the early stages in both the House and Senate — only to have the Administration, for political reasons, reverse itself and actually support the legislation.

More than anything else, the passage of this legislation symbolizes a state of social bewilderment about science and how we should get it done. There are other signals as well: the hot-and-cold recent history of investment in genetic engineering; the disturbing differences of view between the universities and government about federal authority over the transfer of technology; and the controversy over how much corporate support universities should accept for their research programs, and under what conditions. Some changes now underway in US innovation management will be considered now. The argument will be that the social sponsorship of discovery is being rearranged in a very fundamental way. Specifically, I shall try to make the case that we are seeing a revolutionary compression of the innovation process I described earlier; and that, as a consequence of that compression, traditional roles are being redefined and well-understood customs challenged. I will offer some suggestions as to how universities and the scientific community might cope with these changes.

Recent Changes

To begin, let me list some forces that are tending to compress the innovation process and to blur the relationships that up to now have characterized its three stages. I think I can recognize four.

- First, a number of scientific disciplines are now being recognized as "ready" for accelerated application. Perceptions of what is possible are sharpened in such fields: as a discipline matures in power and confidence, leaps from the laboratory to applications that once seemed intimidating become commonplace. That now appears to be the case, for example, in immunology and "genetic engineering" as well as in microelectronics.

- Second, there is a growing social awareness of the importance of scientific discovery to national productivity, and a consequent impatience with the traditional time requirements for technology diffusion. In the past ten years, various studies (for example, that by Comroe and Dripps[3]) have demonstrated — particularly for biomedical research — that the time lag between initial research discovery and practical application is very long, of the order of a decade or more. Thus the message from a small but growing body of 'research on research' is that better devices are needed for speeding technology transfer.

● Third, there is increasing concern in the centers of fundamental research — and by that I mean the research universities, where over two-thirds of the nation's basic science is done — about the retreat in public support of that activity. Federal funds for non-defense research have shrunk by about 38% in real dollar value since 1968, and half that decline took place in the first two years of the 1980s. It is therefore, no surprising that the universities and university scientists are looking for alternative means of support.

● A fourth and final cause, perhaps the most unexpected of all, has to do with the venture-capital financing of small, research-intensive companies in fields like biotechnology and microelectronics. In the nearly four years since major changes were wrought in the capital-gains tax, the investment available for such ventures has jumped from an estimated $70 million in the mid-1970s to about $1.5 billion in 1982. Very large changes in value can take place with successive generations of private investment in high-technology firms, and larger changes still when the firms go public. At the time of its public offering, Genentech stock was valued at $38 per share. Shortly after the offering had been snapped up, it soared to $80. As you will observe, things have come back down to earth, and Genentech, still without a product to sell, was listed a year later at $30 a share. But that story is worth telling, even as a cautionary tale. What made Genentech so appealing was the promise of good ideas, of "breakthroughs." Like other high-technology offerings it made much of its affiliation with first-rate university scientists; in short, it promised to short-circuit the technology transfer process. The lesson — and it made a strong impression in the financial community, despite some subsequent disillusionment about the soundness of biotechnology investment — is that in this new work the big potential is associated with the *early* possession of an idea. The result has been an array of new affiliations between the institutions formerly associated with Stage III innovations and scientists doing Stage I research.

To summarize what has happened, the familiar trajectory of American innovation has become compressed in time. Once there was clear separation between an early, fundamental research phase, publicly supported and characterized by open communication, and a later, applied phase financed by private risk capital and featuring proprietary protection of information. I have tried to analyze some of the forces that have led to this compression: the scientific "readiness" of biotechnology and other fields for a broad range of applications, the need for new funds to replace those lost in the federal withdrawal from the support of basic research, and the style of venture capitalization, which places new premiums on the early capture of intellectual property.

The New Environment: Pressures and Problems

The result is an entirely novel mixture of influences upon university scientists and their institutions. For the university itself, there are new and challenging pressures on investment policy (does the institution go into business with its own faculty?), on technology licensing (should the university license inventions to faculty-led ventures? to their competitors? and, if yes, under what terms?), or research contracts with industry (what restrictions on communication are acceptable, and should there be full disclosure of terms?) and on policies relating to consulting, faculty conflict of interest, and the protection of graduate student interests. But many of the problems are simply not soluble by the institutions alone. For the scientists themselves, and the "invisible colleges" that hold them together in national and international networks, there are special challenges: how much can or should journals or societies guard against the withholding of information and exchange for proprietary reasons? and how much involvement *outside* a faculty member's primary institutional affiliation is appropriate?

I emphasize the role of the professoriate in this last decision particularly because I doubt that *institutional* regulation will be effective or even very influential in establishing the limits. What is thought to be proper by one's community of scholarly peers will eventually set the standards; the best institutions can hope to do is reinforce those social norms, and perhaps to play some useful role in helping to shape them.

Having said that, what are some ways to deal with the compression of innovation and the problems it poses for universities and their faculties? To begin with, we need to examine the traditions that have grown up regarding the institutional obligations of the professoriate, for that is the background against which new proposals will have to be evaluated.

The Universities and Their Faculties

Two resources dominate any consideration of the relationship between the faculty member and the university: time and intellectual property. Established tradition is that institutional purposes have exclusive call on the professional time and energy of faculty members during the period of full-time employment. Three things, however, are worth noting. First, institutional purposes are usually interpreted broadly. As a consequence, a variety of activities that might at first be thought private or even self-serving are regularly judged to be in the interest of the university. Service on peer review panels, advice to local and national governments, and work for scholarly organizations and journals "count" as work for the university — often even when the organizations are for profit and the faculty member's participation is compensated. Few universities have carefully-drafted policies on this; the informal test has usually been that the compensation is light (relative, say, to that for consultation) and that the work serves some interest of the discipline and hence — from the university's vantage point — is *pro bono publico*.

Second, limited "consulting" is permitted. Consulting is a professional service, reimbursed at something like the market value of the faculty member's time. It is permitted on the theory that the faculty member will benefit professionally — as may students, and the educational process — from exposure to applied problems which might otherwise never be encountered. It also happens that in many applied fields valuable literature is held under such tight proprietary control that consulting represents the only route of access to it. The university sometimes has a secondary motive for permitting consulting: it may establish useful relationships between faculty and a source of corporate research support of philanthropy. But the time expended on such activities is limited quite explicitly by most universities, to the extent of about one day per week.[4]

Third, it is important to note that what is being claimed here is *professional* time, not time, period. Neither the university nor the government cares how much "free" time a faculty member spends on remunerative "moonlighting," as long as it is outside the professional role of scholar/teacher—and, of course, as long as it does not infringe on the performance of that role. Thus, paradoxically, it is all right to consult all day Wednesday, but not again on Sunday; whereas it is perfectly permissible to teach skiing on Sunday, but never on Wednesday.

Now that the matter of time is understood, it would be useful to discuss intellectual property. The difficulty here is that different traditions have developed, chiefly out of convenience, to deal with different kinds of property. As to the most basic issue of all—who owns the rights to a discovery made by a faculty member on university-compensated time in a university facility—there is lack of consistency. Federal law vests patent ownership with the inventor. A few universities vest ownership in the faculty member where contracts or grants permit it, but most claim it for the institution, through required patent agreements with faculty members. The difference is not thought significant, or even known, to most faculty members; but to a few it is critical.

Now, created works are different from patentable ideas. They are subject to the laws of copyright, and in this area federal law gives ownership to the institution rather than the individual. Most institutions assert that ownership—even those that do not do so for patentable processes or devices. Thus the professor who owns the rights for a piece of computer hardware does not own the rights to the software developed to go with it. But from this zone of apparently sweeping institutional claim, an important class of exceptions has been carved out by custom. The university does not assert copyright to books written by faculty—scholarly, text, or trade—nor does it try to attach other creative works. Reflected glory, and that alone, is the traditional institutional reward for such faculty *oeuvres*. The issue is not exactly trivial: it is easy to forget, amidst the present excitement, that the universities have sponsored many more millionaire textbook writers than millionaire computerists or cloners.

It is a little difficult to perceive a structural rationale arching over all of this. The difficulty may be pointed a bit if you imagine that you are a space visitor, unfamiliar with our economy or traditions, trying to figure out our rules. After learning the foregoing, you are told that a university employs a surgeon and a sculptor on its faculty. Each has learned extraordinary skills in a demanding course of graduate study; each does campus-based work that is hailed as pioneering; each teaches in credit-bearing courses for advanced students. Which, you are now asked, is allowed to practice for financial return outside the institution? I suggest that you will be unable to answer the question without some special knowledge—consisting, in this case, of the history of clinical practice restrictions imposed by US medical schools to achieve the 'geographic full-time' system, and of some artistic traditions.

It follows that normative judgments are difficult to make in this universe. Limits are placed upon outside commitments with respect to time, not income — except when special considerations intervene, in which case we limit income. We define some work-products of the creative process as belonging to the university, but others as belonging to the scholar. And — as the sculptor/surgeon example shows — we treat even the same kind of work differently in different parts of the same institution.

So — to summarize this account of university rules — there is not a consistent ethical framework for outside involvement by faculty, or for the ownership of intellectual property. Yet there are some guides that may be helpful in dealing with the new incentives resulting from "innovation compression."

Dealing With the New Environment: Some Principles

First, one of the real values in our system of fundamental research is the coexistence of research and training. As my colleague Robert Rosenzweig put it so well in the introductory chapter of *The Research Universities and Their Patrons:*

A research university is one whose mores and practices make it clear that enlarging and disseminating knowledge are equally important activities, and that each is done better when both are done in the same place by the same people.[5]

The university should, in its own interest and society's, pursue policies that act to retain that connection. These will include resisting proprietary influences upon university research, as well as supplying incentives for faculty members to concentrate their own research activities at the university, under university rules. They will also give careful attention to the rights of students: the right of access to information about research programs, and the right of freedom from exploitation in the interest of a faculty member's research program, whether it is on *or* off campus.

Second, the university is entitled to primary claim upon the time and energy of its faculty. That claim can, in practice, only be pressed so hard — as the University of Chicago found to its sorrow, several decades ago, when it tried to retain all forms of professional income, while offering what it claimed to be "full salary." That experiment failed, probably because tradition is powerful and we retain strong instincts about the ownership of private ideas. Still, there are strong arguments in favor of the university's claim. It is paying for full time, in most cases. It does provide the facilities for scholarly work, not to mention the environment. Its institutional *imprimateur* lends value to whatever the faculty member does, because the outside world has learned to respect its judgements about quality. That does not mean that the university owns the faculty member; but it does constitute a primary claim on time and energy.

Third, there is a balancing flow of benefits that dictates institutional restraint in pressing that claim. *Whatever* a faculty member does that is professionally creditworthy — and given the quality of our selection process, that should be the overwhelming majority of his or her output — redounds to the prestige of the institution. That should make us wish to be generous about terms and conditions; after all, if we get the credit anyhow, ownership is perhaps not so important.

Policy Directions for the Universities

What position, given all these considerations, should the university take with respect to outside faculty involvement? I think it should continue to make its primary claim, but permit limited outside involvement when that clearly offers benefits *both* to the faculty member's professional development and to the university-based research and training programs. Disclosure of these relationships should, as now, be required on demand — and, if it proves necessary, as a matter of routine. Management responsibilities outside the university should generally be viewed as inconsistent with the primacy of university commitment — including significant personal responsibility for any regular laboratory operation off campus. Such roles not only become compelling time commitments: they may lead, as we have learned, to awkward mixtures of proprietary and non-proprietary work. They also encourage the introduction of secrecy to university-based programs, and impose troublesome conflicts on graduate students.

On the other hand, I think that intellectual property rights should be vested primarily in the faculty member. After all, one of the attributes of university research that distinguishes it from that done in corporate settings is its independent, individual character. If we are to assert primary claims to faculty time and energy, it would seem appropriate to reward that loyalty by assigning the proprietorship of intellectual accomplishment to the innovator. It is then up to us to structure incentives that will encourage inventors to assign rights to the university and to share the returned value with the entire enterprise.

However the university obtains control over intellectual property, it must consider the responsibilities it accepts along with that control. First, there is an institutional obligation to the process of technology transfer. No valuable invention should be delayed in its progress toward human service by the institution's own need for financial return. In most cases, that principle will lead to a first preference for non-exclusive licensing, and a fee structure designed to encourage broad participation by commercial developers. Second, there should be care to avoid pressures on faculty and students to concentrate unduly upon "developable" ideas, or to keep secrets in the course of developing them. Finally, the university-as-licensor must take careful note of its other roles as investor and research proprietor, and maintain a strictly neutral policy with respect to licensing ventures that may be partly owned by the university or its faculty members.

Even if it has solved all of these problems, the university has still another role: the design and execution of policies regarding the acceptance of research projects. In practice this is usually a matter for decision by members of the research faculty — who resemble, as someone has said, a federation of small, sometimes even medium-sized entrepreneurs. The terms are limited by university rules worked out by faculty committees; but if an outside agency proposes terms unacceptable to the faculty member who is to do the work, there will be no project.

The evolution of institutional policy *and* of a tradition of faculty acceptance has been heavily conditioned by the patterns of federal support that have developed in the postwar years. There are no restrictions on publication, and the federal patrons of research have, as I said, not usually claimed patent rights. By contrast, a number

of private research sponsors have asserted such rights in advance, through requirements that they hold exclusive licenses on any patentable discoveries made
under their support.

Where the institution claims no such rights, the acceptability of such conditions
becomes a matter for the faculty member to determine. But where the institution
retains rights to the intellectual property, the decision becomes more complex and
difficult. If a faculty member were to assert that a line of important work could only
be supported *if* a private sponsor were granted exclusivity in advance, I think the
university would agree to take the project. What if the institution were deciding
the matter on its own? I confess I would have difficulty. The situation offers too
many temptations for *a priori* restraints on communication, too much emphasis on
investment return, to leave me comfortable. I expect this to become one of the
main policy issues in this area, the more so because large grants of this type have
been made to several institutions.

The university does, of course, impose rules designed to protect freedom of access, publication rights, and other features that might be vulnerable where there is
strong proprietary interest. In many places these are being reviewed to see if they
are strong enough. It is important to note that this is *not* entirely, or even mainly, a
response to the proprietary impulse. For entirely different reasons the government
has moved toward the application, in certain research contracts, of restraints upon
access or freedom to publish promptly. Because we clearly cannot have one set of
rules for industrial research sponsors and another set for the government, these
federal actions have added urgency to our consideration of the value of openness in
research programs.

The case usually made for freedom in scientific communication rests on the need
for rapid transfer of methodologies and ideas from one research group to another,
and on the demonstrated utility of published and pre-publication exchange in
maintaining the vigor of a discipline. It derives mostly from convincing assertions
by working scientists, nearly all of whom say it is essential. A few more formal
studies also support the special role played by informal communication in advancing a new idea; notable among these is the analysis by Wolff of the exchange of
pre-publication information in the development of our understanding of the role
of cyclic AMP as a cellular "second messenger" — a function Wolff wittily attributed
to informal exchange among scientists in the title of her paper.[6] I think it is safe to
say that both that kind of communication — ranging from laboratory visitation between scientists to attendance at scientific meetings — and the more formal sort en-
tailing publication are critical parts of the scientific process.

There is a second part of the case that relates only to formal communication.
Science relies very heavily on the capacity to replicate experiments; it is the only
systematic way to recognize error, and the only way at all to correct fraud.
Although we referee journal articles, evaluate the logic of propositions, and check
arithmetic, we cannot decide, merely from reading a report, that a result is
right — only that it is not wrong in some obvious way. Accordingly, we require that
scientific communications include enough detail about the way an experiment was
done so that a competent investigator in the field can do it in exactly the same way.
This is an exacting requirement; it compels the release of *all* relevant information

about methods and techniques. Secret ingredients, magic sauces and your own special glassware are fine in cookery or in product development, but in fundamental science they are *out*.

Proprietary and governmental threats have forced universities to take some recent actions to protect both kinds of communication, and it is not unlikely that the challenge will escalate. Using the regulatory authority provided by the International Traffic in Arms legislation, the Departments of Commerce and Defense have attempted to restrict US scientists (and succeeded, in some cases) from delivering papers at international meetings on basic research relating to "critical technologies." Language has been attached to university research contracts that would restrict the access of foreign graduate students or fellows or even faculty members to university projects. And the Department of State has sought to place onerous monitoring requirements on universities receiving Soviet bloc visitors so that the latter will not have access to projects or even conversations on critical topics. We have argued vigorously against these restrictions, sometimes refusing visitors or contracts; and we are trying hard to press upon the government our own conviction that the social value inherent in open science far outweighs the losses we may risk by being too permissive about the transfer of technologies.

As to proprietary influences, most universities—as I have said—refuse any but the briefest delays in publication time, and reject restrictions upon access to programs. They have also had to cope with the increasing threats to informal communication posed by the actions of commercial firms, especially in biotechnology. Several active or prospective litigations concern the patenting of DNA sequences by companies who obtained required material through informal exchanges with academic scientists. And for each lawsuit, there are a dozen horror stories. Whatever the merits of these, the perception is that people are prepared to take advantage of naive informality on the part of academic researchers. The latter, predictably, are reacting by becoming more cautious. Meanwhile the universities are trying to provide them with increasingly aggressive legal protection, and evolving policies for the distribution of research-derived materials that will prevent their commercial appropriation while encouraging academic circulation.

Policy Directions for the Scientists

In this as in other areas, however, there are sharp limitations on what the universities can accomplish themselves. I began this section with an assertion that the compression in the course of innovation and the consequential reordering of how and where it is accomplished requires much more than the development of new regulatory policies by universities and new patterns of financial support by government and industry. It also challenges the scientific community itself to develop new rules, because many of the current problems are beyond the regulatory reach of the institutions. Although many people suppose that something vaguely called "academia" has a single position or set of attitudes on such matters, that is actually far from the truth. Frequently the interests of working faculty scientists and the university science administrators who are their former colleagues diverge sharply; indeed, one can now regularly find them giving conflicting testimony before the

Congress on such issues as indirect costs and mandatory retirement. In the present case, their interests are not so much divergent as different, and complementary.

Standards of the profession, publication policies, criteria for the acceptance of work, and peer review are functions that the professoriate rather than the academy must regulate. And there are other areas in which regulatory controls by the university, though theoretically plausible and often urged, would probably prove unacceptably corrosive in practice. Extreme forms of regulation of outside involvement, or Draconian limitations on faculty ownership of intellectual property, would be examples. Thus there are two reasons for the assertion of self-regulation by the scientists and their networks. The first is because the task is more appropriate to them; the second is that failure to do so may invite unwelcome attention from others. It should be pointed out that regulatory attention by the *university* is not the only, and not even the worst, alternative prospect. The Congress is now making concerned gestures about matters ranging from the treatment of animals to scientific fraud, and one hears proposals for government auditing of research to insure proper practice.

What areas are particulary important for self-regulation by scientists? One surely is the firm protection they can give to the open character of scientific work. Journals should not publish any articles from which proprietary information has been held back so as to make the work more difficult to replicate, and societies should refuse to allow papers to be presented at meetings if they are subject to similar restraints. Individual scientists can reject efforts to attach access restrictions or similar provisions to research grants and contracts. In these ways, the scientific community can make clear the essentiality of open communication to the conduct of its affairs.

A second area requiring self-regulation involves the public perception that fundamental science is now "commercial," and that its practitioners are subject to temptations that cast doubt upon their objectivity. This is a serious problem, which has been unduly abetted by some biomedical scientists, who have reacted to financial opportunity with an appalling lack of restraint. The use of university facilities to house the back-contracted spillover from an outside commercial venture; the use of a university position and laboratory to promote corporate commercial claims for a new venture; the coercive inducement of graduate students to "join in" on an outside operation: all these have happened at major universities within the past three years. Surprisingly, there has been little criticism of the researchers involved from within the scientific community. I find that shocking, given the willingness of scientists to engage in frank criticism of institutional practice in these same areas. Unless there is firm rejection of such behavior, the public will be perfectly entitled to conclude that the scientists are more interested in profits than results. Fundamental research is still deeply respected in this country, much more so than most professions. Public confidence in its own objectivity is the best thing science has going for it; but without some care we may find one day that "scientific objectivity" has become an oxymoron, a self-contradiction like "express mail" or "easy credit."

Notes

1. Proponents of the legislation argued that funds for the small business innovation programs need not be set aside from basic research programs. But the law permits a diversion of basic research funds to product innovation, and that diversion will surely happen to some extent. Proponents also claim that small businesses are capable of, and want to conduct, truly basic research—a proposition that few scientists would accept.

2. There were some early opponents, notably the American Association of Medical Colleges, but for the most part the political arms of the university and scientific organizations did not become actively involved until late in the legislative process.

3. Julius H. Comroe, Jr., and Robert D. Dripps, "Scientific Approach to a National Biomedical Science Policy," *Science* 192 (1976), pp. 105–111.

4. It is often asked, of course, whether "weekends count." Some universities aren't prepared to say, but government rules about professional effort establish a *de facto* policy—and under it, weekends most assuredly *do* count. That is, all of the time one spends professionally, whether it is 40 hours per week or 100, constitute "100% time." Accordingly, a faculty member who has a government grant and consults Friday, Saturday and Sunday is accounted as spending 3/7ths time consulting—which is against government rules if that faculty member is reporting the usual 100% research effort on the grant. So in practice one day a week means just that.

5. R.M. Rosenzweig and B. Turlington, *The Research Universities and Their Patrons* (Berkeley: University of California Press, 1982).

6. Patricia Woolf, "The Second Messenger: Informal Communication in Cyclic AMP Research," *Minerva* XIII:3 (1975).

5

Is the Biological Revolution a Match for the Trinity of Despair?

Richard M. Krause

Biology has been seized by a revolutionary fever. The word has spread through the popular press. Even Wall Street trades in issues that capitalize on the new biology. What is this so-called "revolution," this new biology, all about? One thing is clear: biology as we knew it will never be the same again. The turbulence of this revolution will spill over into our daily lives. There is no escape.

Perhaps some of the readers of this Journal studied the old biology, as the author did before this revolution occurred. Biology was an idyllic exercise in those days, a refuge for professional and amateur alike. Vladimir Nabokov, the emigre Russian novelist, for example, was an authority on butterflies. There was nothing very practical about our observations. Except for those who went into medicine, biology was all rather pointless, in comparison to physics and chemistry. But that, of course, was its charm.

Analysis of the world around us requires the application of mathematics and the physical sciences; often these were the very sciences we sought to avoid as we flocked into freshman biology. But all that has changed. The biological revolution has swept away for all time this leisurely examination of living things.

The signpost outside the door of the biology lecture room and laboratory of today bears this warning: "Do not enter without knowledge of chemistry, physics, mathematics and computers. Without these tools you will lose your way." At the core of this new biology is a highly complex process called recombinant DNA technology. What is most remarkable about this new development is not the effect it is having on every nook and cranny of biology and medicine, but the practical applications of this technology in medicine, agriculture and industry.

We are promised new vaccines, new miracle drugs, cures for diseases that still elude us. Nothing alive will escape this new influence. Some predict that agriculture will be revitalized, that sturdy crops more resistant to insects can be developed, that perhaps even healthier breeds of livestock can be raised, that new sources of clean energy and new methods to control environmental pollution will be developed—I could go on and on.

To be sure there is, as with all revolutions, alarm in some quarters: fear that this new biology will alter the shape of living things, perhaps divert the stream of evolution, take away from the Almighty himself the authority to blaze the evolutionary trail. Others have cautioned that such an evil influence might even extend to human reproduction, generating clones of people with predetermined characteristics: glamor gals in abundance, or bulging muscle men, produced with nary a barbell or Nautilus machine.

Most surprising, perhaps, is the influence of this new biology on the marketplace. On the speculation that drugs developed by the new biology may cure cancer, prolong life, and provide health, venture capital has launched over 300 new companies specializing in the practical applications of recombinant DNA technology. Their names are rapidly becoming familiar to investment brokers: Genentech, Cetus, Genex, and so on. In 1981 Cetus attracted $100 million in venture capital from major corporations: All of this stirred up by the enthusiasm of a public always eager for the elixir of youth and the power to turn base metal into gold.

But a cautionary note: Despite all the enthusiasm, there has been little financial return so far, mostly in speculative profits. Clearly the immediate promise has been oversold, but if all goes well, the long-term potential staggers the imagination. And so, despite the risk, the driving force behind the eagerness of venture capitalists is, in Samuel Johnson's words, "the potentiality of growing rich beyond the dreams of avarice."

"The Trinity of Despair"

But enough of the promise of riches. Will the biological revolution be up to the major tasks before us? Will it ease the burdens of humanity? Will it heal the sick, feed the hungry, clothe the naked? Will it solve the ills that afflict this crowded globe, the ills I have called the "trinity of despair"? It seems to me that these are the overriding questions. And so I ask the question: Is the biological revolution a match for the trinity of despair? Let us examine the elements of this trinity, one at a time.

Hunger

For the masses of the world, hunger stalks the globe. For many, it is more than hunger. For many, it is the fear of famine and starvation. Each day one-third of all

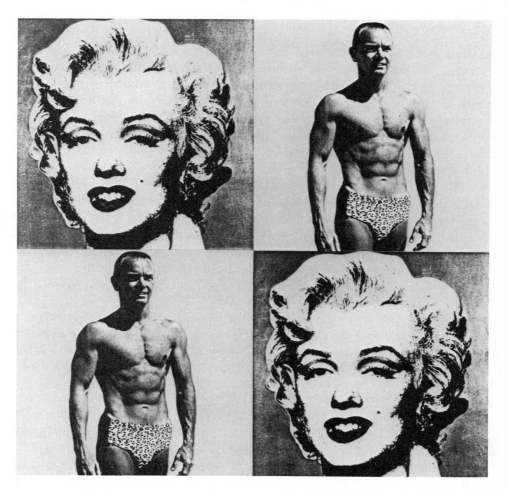

FIGURE 1. A combination of Andy Warhol's portrait of Marilyn Monroe and a photograph of an anonymous "muscle man" in the style of Andy Warhol.

men, women and children in the world have a substandard diet — a diet that does not meet minimal nutritional requirements. And, beyond hunger, five million people are starving.

Disease

We must contend not only with major diseases here at home — diseases such as cancer, heart disease, diabetes — but with less familiar diseases in faraway places — in Asia, in Africa, in South and Central America. I speak of debilitating diseases that flog the body and drench the spirit; parasitic diseases with strange-sounding names and still others not so strange. Malaria, for example, was common in Georgia until a few decades ago. And today, in much of the world, malaria, "the king of diseases," still saps the strength and energy of over 200 million of the world's poor.

Population and Supply

The third member of the trinity is somewhat more difficult to define. The third element of this dismal group arises from insufficient resources to support an expanding population. The people of the world — growing in numbers — must scramble to accommodate human needs to the limits of the earth's resources. Demographers and economists are gripped by a Malthusian gloom as they contemplate a projected world population increasing at an exponential rate from four billion today to over six billion by the year 2000. Can we provide resources for all of these additional people?

We are living in the "twenty-ninth day," as Lester Brown puts it in his book of the same title. The French use a riddle, he notes, to teach school children the nature of exponential growth: A lily pond contains a single leaf. Each day the number of leaves doubles: two leaves the second day, four the third, eight the fourth, and so on. If the lily pond is full on the thirtieth day, the question goes, at what point is it only half full? The answer, of course, is on the twenty-ninth day. Our global lily pond on which four billion of us live is already half full. We are living in the twenty-ninth day.

So there you have them — the three members of the trinity — hunger, disease, and the limited ability of the earth's resources to support the exponential growth of population. Is the biological revolution a match for the trinity of despair? Let us now investigate this question.

The Biological Revolution

The origins of the biological revolution began over 100 years ago, when Darwin developed his theory of evolution and Mendel, a contemporary, performed his pioneering work in genetics. The full fury of this revolution erupted in the 1950s with the discovery of the genetic code. Since then we have been propelled forward with ever-increasing frenzy, and the tempo has been at full throttle with the recent introduction of recombinant DNA technology. As with all revolutions, we must trace its origins if we are to comprehend its influence on our daily lives.

Gregor Mendel, who lived and worked over 100 years ago, was the grandfather of modern genetics. Mendel was a remarkable man, a man of the cloth, ecclesiastical. He was a scholar in a period when knowledge and theory were interwoven, a fabric that — in this century — has become unravelled.

Mendel's brilliance was his recognition that inheritance was determined by a memory bank of particles we now call genes. These genes contain instructions, or blueprints, for making all internal components of the body, as well as such externals and recognizable personal characteristics as size, shape, hair color and eye color.

Early in this century, scientists suspected that perhaps it was the cell nucleus that contained the specialized genetic material. Later, cytologists observed that, during certain phases of the life cycle of the cell, the material in the nucleus was arranged in rod-like structures, called chromosomes. During the cell division, these chromosomes are duplicated, doubling in number and forming pairs. The chromosome pairs part company and one copy of every chromosome is distributed

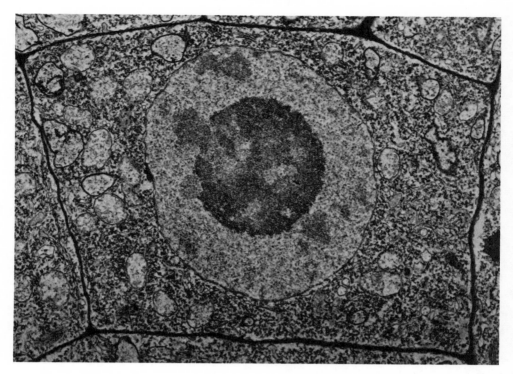

FIGURE 2. This electronmicrograph shows the rectangular structure of a typical human cell. The round central nucleus contains millions of genes which carry the instructions for making all the body's components.

to each of the two daughter cells that emerge from the original parent cell. In this way, all of the genes—all the genetic information that was present in the initial cell—is passed on to the daughter cells. This process is repeated millions of times a day in each of us to replace worn-out red blood cells, dead skin, and —in men—millions of new sperm, restoring the supply released through intermittent discharge of semen.

From a long series of complex studies emerged a theory that genes were particles of information strung out in linear array on the chromosomes, much as a collection of colored beads on a string. But what the gene was and how it carried instructions for red hair or blue eyes, the beauty of Marilyn Monroe, or the bulging muscles of an Arnold Schwarzenegger, remained a mystery for some time to come.

Around 1940 the first tremors of the biological revolution were felt as scientists began to uncover the secrets in the genes. They found that genes contain a code that, once translated, dictates precise instructions for the development of red hair or blue eyes as well as all other individual characteristics. They reasoned that, because offspring acquire duplicates of parental genes, the code is transmitted from one generation to the next; for this reason, there is a resemblance between parents and offspring.

We cannot review here the long series of observations that, in time, revealed the

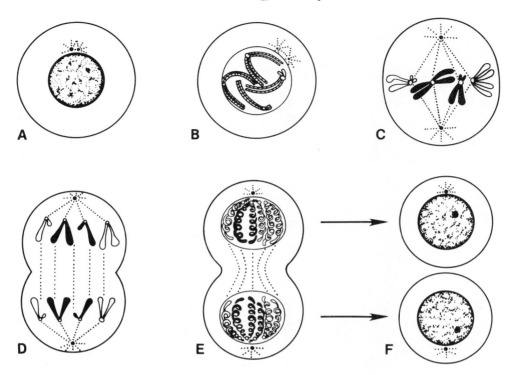

FIGURE 3. Cell Division: a) a parent cell in normal functional state; b) duplication of chromosomes; c) alignment of chromosomes in preparation for equal distribution; d) equal distribution of chromosomes; e) repackaging each chromosome set; f) separation into two identical daughter cells, each possessing an exact replica of the chromosomes in the parent cell.

nature of the genetic code. But one critical discovery was made in 1944 by Oswald T. Avery, Colin MacLeod, and Maclyn McCarty when they found that the genes consisted of a strange and mysterious chemical substance called deoxyribonucleic acid (DNA, for short), and that DNA is the carrier of the genetic code. Scientists also found that DNA contained four different chemical bases. These bases are named adenine, guanine, cytosine, and thymine, or "A," "G," "C," and "T" for short.

In 1953 Watson and Crick electrified biology when they proposed that DNA consisted of a double helix, rather like a long, spiral staircase with the four bases, in various sequences, forming the steps of the staircase. Some suggested that the bases were like the letters in the alphabet, arranged to spell code words. But what were the words and what did they mean? Were these, perhaps, the code words that guide the chemical machinery of the cell to produce the proteins that mean life?

One of the difficulties in answering these questions was that there seemed, at first, to be too few letters in the genetic code, too few to code for all the complexities of life itself. After all, our own English alphabet has 26 letters. Surely if 26 letters are needed to write a Shakespearian sonnet, how could an alphabet of only four letters spell out instructions for all the complexities of life itself? In time, the riddle was solved, and this is how it happened.

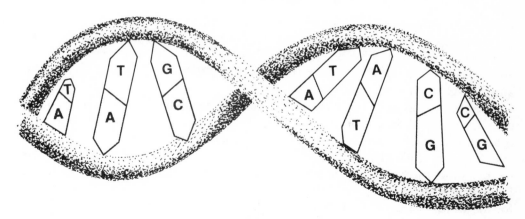

FIGURE 4. DNA consists of a double helix, like a long spiral staircase with various arrangements of the four bases forming the steps.

One fact was clear. The genes consisted of only DNA. Thus, scientists reasoned that somehow the four chemical bases in DNA must code for 20 amino acids, constituents of the hundreds of proteins necessary for life (see Table 1). Furthermore, these proteins can consist of any number of amino acids strung out in complex combinations. Two questions, then, were at the very heart of the matter. How could only *four* chemical bases code for 20 amino acids and how do these amino acids combine to build hundreds of different proteins?

To explain how this mystery was solved, let us create an analogy. We will think of the nucleic acid bases as letters of an alphabet, amino acids as words, and proteins as sentences. At first, scientists thought that perhaps the "words" for amino acids could consist of different numbers of letters, just as the words on this page consist of different numbers of letters. But, surprisingly, this was not the case. What nature did was to use only three-letter words for the 20 different amino acids, wisely steering a wide path around all four-letter words. (Maybe the Almighty really does have his hand on the tiller!)

A single word of three nucleic acid bases, called a codon, specifies one amino acid. The four bases can be arranged in 64 different combinations—64 different words, if you will—more than enough to represent the 20 amino acids. In fact, just

TABLE 1. DNA Chemical Bases and Their Amino Acid Constituents

Nucleotide Bases in DNA	Amino Acids in Proteins		
Adenine	alanine	glycine	proline
Thymine	arginine	histidine	serine
Cytosine	asparagine	isoleucine	threonine
Guanine	aspartic acid	leucine	tryptophan
	cysteine	lysine	tyrosine
	glutamine	methionine	valine
	glutamic acid	phenylalanine	

as there are synonyms in English, there are several different combinations of three bases that specify the same amino acid. For example, in the DNA alphabet, the three-letter combination of bases C-A-T does not spell "cat" at all, but rather the amino acid *valine*. In genetic language, T-A-C spells the amino acid *methionine* and T-A-G spells *isoleucine*.

There is not space here to review the long and arduous task of breaking the genetic code, but it was, in many respects, similar to breaking the hieroglyphic code on the Rosetta stone, a slab bearing parallel inscriptions in Greek, Egyptian and ancient hieroglyphics. The Frenchman Champollion deciphered the hieroglyphics by comparing the letters in the ancient hieroglyphic alphabet to the letters of the Greek alphabet. A comparison of the hieroglyphic and Greek symbols for the words "Ptolmis" and "Kleopatra" broke the code, or rather translated the ancient language to the more modern Greek.

Now that we have seen how DNA "words" code for amino acids, let us answer the second question: how are amino acids arranged to build a protein? Human insulin is a protein hormone that helps regulate glucose metabolism. The structure of insulin is a "sentence," made of 17 different amino acid "words," some of them used more than once in an English sentence. There are, therefore, a total of 51 amino

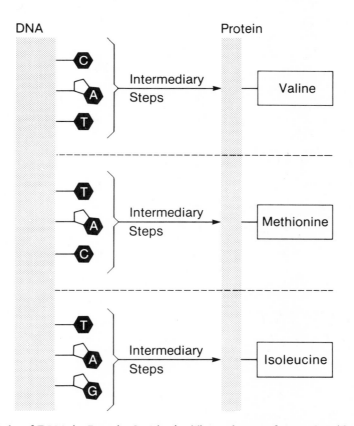

FIGURE 5. Role of DNA in Protein Synthesis. These three codons—C-A-T, T-A-C, and T-A-G—"spell" specific amino acids and direct the cell's intermediary machinery to synthesize a portion of a protein consisting of valine, methionine and isoleucine.

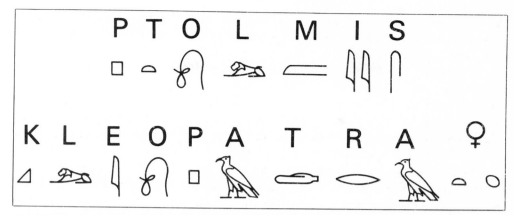

FIGURE 6. A letter-by-letter comparison of the words "Ptolmis" and "Kleopatra" with the hieroglyphic symbols for the same two words. Note the identical symbols for "P," "O" and "L" in the two words. In "Kleopatra" the letter "a" appears twice; therefore, Horus the hawk must represent the letter "a." The half-moon and oval symbols at the end of "Kleopatra" are the feminine termination.

acids in this sequence. In the DNA that codes for insulin, there are 51 three-letter words, called codons, arranged in the same sequence on the ribbon of DNA as are the amino acids in the protein insulin.

Think of DNA as a long ribbon. In fact, although the cell itself is only 100,000th of a meter in diameter, the DNA coiled up within the nucleus of each cell is one meter (39 inches) long. (Think of it another way: If the individual cell were one foot in diameter, the ribbon of DNA would be 20 miles long.) All along this one-meter length of double helix are clusters of three-letter code words forming the "sentences" that spell the many proteins needed for the function of that particular cell. One cluster of three-letter code words might spell hemoglobin, the special protein in red blood cells that carries oxygen to the tissues. Another cluster spells myosin, the protein that makes up the bulk of Arnold Schwarzenegger's muscles. In other words, Arnold has myosin in spades! He has, like you and me, only *one* gene for myosin. The reason he has such large muscles is that, unlike you and me, he has another gene that instructs the body's chemistry to make extra large amounts of myosin.

We said before that the chromosomes containing the genes double during cell division and each daughter cell gets an exact copy. This ability of DNA to duplicate itself is a very important property. It is the process by which human characteristics are transmitted from generation to generation. Biology speaks a universal language. The same voice that transmits the genetic code in humans is used by the most primitive forms of life as well. Indeed, as the Bible says, "the voice of the turtle will be heard throughout the land."

Recombinant DNA Technology

Now that we have covered the basics, let us see how this knowledge has been used to create the new recombinant DNA technology. From a practical point of view, the purpose of this technology is to alter the genetic make-up of certain bacteria so that they are programmed to produce a wide variety of useful pharmaceutical and industrial substances. "Why use bacteria?" you may ask. For starters, we domesticated bacteria centuries ago. Most bacteria are among our best friends. For at least 6,000 years we have used them to make alcohol. (Ancient Egyptian hieroglyphics recorded vintage years for wine and beer.) Since 1930 bacteria have been used to produce vitamins commercially. But bacteria do not contain the genetic information necessary to make many medical, industrial and agricultural substances useful to man: human insulin to treat diabetes, for example, and promising new experimental drugs, such as interferon.

How are bacteria altered using this new technology? Recombinant DNA technology takes advantage of a special feature in the sex life of bacteria. Strange as it may seem, bacteria have a sex life similar to that of all other species. Take away the romance and sex boils down to the simple transfer of DNA. Like humans and all other living things, bacteria of the same species also have a mechanism to transfer DNA from one individual to another.

One form of bacterial sex life concerns special genetic elements called plasmids. This is how it works. The genetic content of bacteria consists of a long strand of DNA and several small, circular bits of DNA, not connected in any way to the long major thread. Called plasmids, certain of these circular bits function as sexual organs that bacteria use to engage in their own special form of copulation. Bacteria are constantly passing these plasmids between each other—a sort of primitive sexual promiscuity that is frequently of advantage to the bacterium. For example, this plasmid contains the genetic power to make the bacterium resistant to antibiotics, such as penicillin. In nature these antibiotic-resistant plasmids are transmitted from bacterium to bacterium. In doing so, bacteria that possess antibiotic resistance can transfer this power to other organisms that do not. These become the so-called "super bacteria" that are resistant to penicillin. It is in this way that bacteria create a whole new generation of resistant bacteria and elude our best efforts to contain them. They survive by swimming with, not against, the evolutionary stream, a principle we modern scientists have far too often ignored. As a consequence, bacteria still outwit us.

Using recombinant DNA techniques we can use these plasmids to insert DNA from any species of our choice into domesticated bacteria: from man, from mice, from plants, from any living source. We can be very particular about the characteristics of this alien DNA and, in this way, control exactly what new quality we add to the bacteria. Our strategy, of course, is to use the DNA code words that will instruct the bacteria to make substances of use to us: new drugs, new vaccines, new hormones.

This is how it is done. Foreign DNA—the piece of DNA that "spells" insulin, for example—is cut from the nucleus of a human pancreas cell, where insulin is formed, and inserted into a plasmid from *E. coli,* a domesticated bacterium. Remember that DNA reproduces itself during cell division, a process that occurs every 30 minutes in bacteria. Each daughter cell now has three plasmids and each plasmid contains that additional bit of human DNA that codes for the word insulin. As these bacteria grow, insulin is produced and accumulates in a laboratory vessel con-

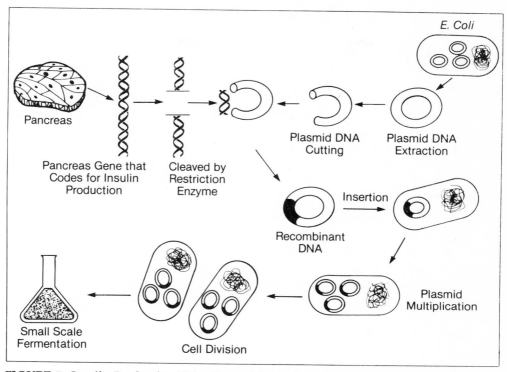

FIGURE 7. Insulin Production Using Recombinant DNA Technique. Using recombinant DNA technology a segment can be cleaved from a piece of DNA from humans or animals and inserted into plasmids of *E. coli,* the most commonly used domesticated bacteria. In this example, recombinant DNA techniques are used to produce human insulin in the laboratory.

taining the bacteria. It is then purified and prepared for human use. This has been done, in fact, by the Eli Lilly Company.

From a theoretical point of view, there is no end to the potential applications of this technique to the manufacturers of other substances useful to humankind.

A Solution to "The Trinity of Despair"?

But what of the trinity of despair: disease, hunger and the mismatch between dwindling resources and an expanding population? Will the biological revolution truly provide solutions to these problems?

Disease

I could, of course, tell of how the biological revolution may cure cancer and even the common cold. But it seems to me that we should cast a wider net and examine several important health issues of the world at large. "The whole earth is our hospital," observed T.S. Eliot and, indeed, it is.

I already mentioned malaria as a major illness that afflicts several millions of the world's poor, particularly in tropical Africa and Asia. Of all the parasitic diseases, malaria is not only the greatest killer, but the most widespread disease in the world.

The human malaria parasite develops alternately in two hosts, in humans and in the anopheles mosquito. Approximately 10 to 25 days after a susceptible person is bitten by an infected mosquito, the parasites — which mature in the liver — invade the red blood cells and multiply. This blood infection produces the symptoms of the disease: chills, fever and profuse sweating. With this long, lingering illness, patients cannot perform a full day's work, especially the hard manual labor frequently required of the residents in these parts of the world.

We have used many public health measures to prevent the spread of malaria: control of mosquitoes by use of insecticides, such as DDT, and the draining of swamps where mosquitoes breed. In addition, modern medicine has provided chloroquine for treatment. But evolution marches on. Through genetic mutations the mosquitoes have become resistant to DDT, and, for the same reason, the malaria parasite is becoming resistant to chloroquine. As a consequence, there has been a resurgence of malaria in Africa and Asia in the past decade. Indeed, in Ceylon in 1970, there was hardly a case, but today there are a million people with the disease.

Yet all is not lost. Because of the biological revolution, there is now the prospect of a malaria vaccine. Had you asked me five years ago about the possibility of such a vaccine, I would have been pessimistic. We had no idea how to develop one. But all that has changed, and for the first time there is hope that malaria — and many other of the world's devastating diseases — will be conquered for good.

Hunger

The biotechnology boom will have far-reaching effects in agriculture, offering hope of feeding those who are hungry. *Science* magazine reported (September 18, 1981), in the article entitled "Biotechnology Boom Reaches Agriculture" that major corporations and a phalanx of new research firms will make a major impact down on the farm. Through genetic engineering, it might be possible, for example, to increase the nitrogen-fixing abilities of the bacteria that inhabit the nodules on the roots of legumes like soybeans. Or recombinant DNA techniques might be used to introduce nitrogen-fixing capacities into plants, such as cereals, reducing their need for fertilizer.

Many young scientists are joining this botanical revolution. To old agricultural problems they are posing new practical solutions, solutions such as introducing into a plant, by recombinant DNA technology, the ability to produce a toxin that would kill an insect predator or a toxin to fight off the effects of fungal diseases, thus increasing crop yields. In fact, one young scientist, Dr. Meagher of the University of Georgia, says, "It's not a question of *if* we'll get there, but of *when* and *how*."

Commercial labs, too, are digging into agricultural research. Bacteriologist Winston Brill, a professor at the University of Wisconsin and an associate of the Cetus Laboratories in Madison, Wisconsin, has already developed soybean plants 50% larger than average. Brill has also crossed domestic corn with rare tropical strains. The root material of the resulting hybrid is capable of supporting nitrogen-fixing bacteria.

In short, we firmly expect that biotechnology will ultimately enhance crop yields. A private study this year put the agricultural potential of the new technology at $50 to $100 billion a year, including the development of improved crops and breeds of animals. In the end, biotechnology may make a bigger difference in agriculture than in the pharmaceutical industry.

Insufficient Resources

But what of the third member of our dismal trinity: insufficient resources to support the growing world populations? This is a complex issue, involving as it does agricultural and industrial practices, world trade in goods and services, a mosaic of contending political philosophies, and even social, moral and religious convictions about such issues as human fertility and the use of birth control to curtail the growth of populations.

It is not my intention here to get caught up in the matrix of the social and political issues concerning birth control. There are, I know, many who believe that vigorous birth control measures are the only solution to the rising tide of the world's population as we pass beyond the "twenty-ninth day." I recall hearing (during a round table discussion 20 years ago) Arnold Toynbee vigorously arguing the need for birth control. He got quite worked up over the matter and—flushed of face and in full voice—explained that we must instruct our young people in the use of birth control methods. As he warmed up to the matter, he went even further: we must insist, he said, we must demand their use of birth control devices! Governor Adlai Stevenson, another member of the round table discussion, was listening intently to Dr. Toynbee, and one could observe the amusement gradually building up on the contours of his face. Finally, Governor Stevenson broke into the discussion and asked, "Professor Toynbee, are you speaking about procreation or recreation?"

Like Stevenson, of course, I must not let humor substitute for a reasoned analysis of a complex issue. As Woody Allen observed, "When you do comedy, you're not sitting at the grown-ups' table."

Population Growth and Grain Production

To analyze the third member of the trinity, let us focus on two aspects: population growth and the yearly worldwide production of grain to feed this expanding population. The balance between demand (people) and supply (grain), now and in the future, is at the heart of the matter.

During the period between 1950 and 1980, the world population nearly doubled, increasing from 2.5 to 4.2 billion. But, because the world's farmers doubled grain production from 631 million metric tons to 1,237 million metric tons, we have maintained a reasonable balance. At first one might take considerable comfort in these data; we really seem to be doing pretty well. A closer inspection, however, reveals the storm clouds gathering in the distance. All is not peaches and cream down on the farm. In a November 1981 issue of *Science,* Lester Brown examines why.

What has, in large measure, sustained the level of our worldwide grain production has been the great American breadbasket. More than 100 countries import North American grain. In Brown's words, "The worldwide movement of countries from export to import status is a much travelled one-way street. The reasons vary but the tide is strong: no country has gone against it since World War II. Literally scores of countries have become food importers but not one major new exporter has emerged."

In a world where 70 million new people are added each year, increased food production can be achieved by either increasing productivity per acre or by increasing the number of acres under cultivation, or a combination of both. Will the gains of the past persist into the future? Clearly, if we are to meet future demand, we will need more efficient and productive agricultural methods. At the same time we will need to mobilize every cultivatable acre. Is the expanding population outstripping the total land available for cereal production? Unfortunately, there simply are limits to the additional land that can be brought into cultivation.

There is one aspect of this problem where medicine may be of help. Vast regions of fertile land in tropical countries may be abandoned because of unhealthy conditions. In Africa, a disease called onchocerciasis, or river blindness, has driven thousands of farmers from fertile valleys. Onchocerciasis is a parasitic disease transmitted by peculiar black flies that inhabit these regions of Africa. A serious complication of the disease is blindness.

Medicine and agriculture must work hand in hand to restore these lands to cultivation. I believe that through modern biology we will learn to prevent this

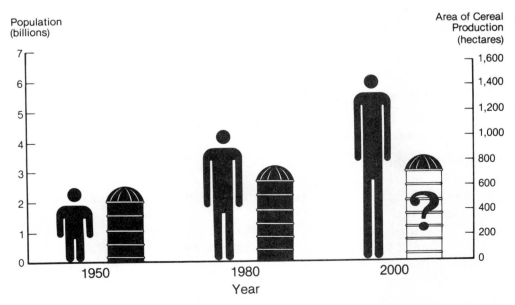

FIGURE 8. World Cereal Production/World Population. Can we again increase the yield of grain by the year 2000 to meet our exponential growth in population?

disastrous illness, and make these regions fit for habitation and cultivation. Already there is promise that ways can be found to eliminate the flies that transmit the disease. But new treatment, new prevention measures, new drugs and new vaccines will take the commitment of many bright young molecular biologists.

Unfortunately, there is frequently insufficient incentive for biologists to make a commitment to these seemingly intractable health problems in faraway places. The "glamor" of research on heart disease, cancer, and diabetes is more attractive. And yet it is in these faraway places with strange-sounding names that we must come to grips with these matters of health for all and food for all. Surely the world will not *remain* a stable place for long if North America is the primary source of grain for an expanding population too sick of body to feed and clothe itself.

The importance of research in tropical medicine relates not only to issues of health — important though these are — but also to considerations of our international commitments and national security. Political instability emerges in the developing countries because the people are consumed by the trinity of despair. The tide of this despair will swell upon our shores and will not recede until we recognize that health for all means wealth for all.

The Moral Responsibility

Setting aside such pragmatism concerning the security of the United States, I cannot close without raising issues of our moral responsibility to assist the poor, the hungry, and the sick.

Two years ago I attended a play entitled *Semmelweiss* by Harold Sackler. The play told of the 19th century Viennese physician, Ignaz Semmelweis, who proved that pregnant women in the delivery ward contracted a lethal infection called "childbed fever" from physicians whose hands were contaminated with deadly germs. Childbearing was a risky business in those days.

The play closed after two weeks, but Semmelweis, the man, deserves our attention. Through the eyes of Sackler, poet and playwright, the significance of Semmelweis's discovery was endowed with a special humanity. Through careful observation of many childbirths, Semmelweis concluded that the hands of the physicians, unlike those of the midwives, were contaminated with germs from autopsy material. When, in some cases, Semmelweis convinced his colleagues to disinfect their hands before they moved from autopsy room to maternity ward, their patients were spared the lethal infection. But these convincing observations were greeted with doubt and skepticism. Semmelweis's advice was not accepted by the medical establishment, those "gravediggers with forceps," as Samuel Beckett calls them in *Waiting for Godot*. Semmelweis was discredited. While doing laboratory research on childbed fever, he died of the same infection that caused this illness. The tragedy of young Dr. Semmelweis is humanity's tragedy. And so the play is not a success story, but a morality play.

Writing about his play, Sackler said, "we stand in the face of suffering and death, ultimately helpless. Yet the measure of man is the degree to which he refuses to live by this commonplace. Somehow to intervene, even briefly, between our fellow creatures and their suffering or death, is our most authentic answer to the question of our humanity."

Conclusion

Is the biological revolution a match for the trinity of despair? You bet it is! Surely there has never been a time when we were so well placed to match our wits with the unpredictable events of history. As George Santayana, the 19th century Spanish philosopher and poet, put it, "those who cannot remember the past are condemned to repeat it." Will we survive beyond the "twenty-ninth day"? Will disease and famine overtake us? Let us profit from history. If we learn to use the power in the biological revolution wisely, I do believe that we can reach the promised land.

Bibliography

Avery, O.T., C.M. MacLeod and M. McCarty, "Studies on the Chemical Nature of the Substance Inducing Transformation of Pneumococcal Types," *Journal of Experimental Medicine*, no. 179 (1944).

Brown, L.R., *The Twenty-Ninth Day: Accommodating Human Needs and Numbers to the Earth's Resources* (New York: W.W. Norton, 1978).

Brown, L.R., "World Population Growth, Soil Erosion, and Food Security," *Science*, November 27, 1981.

Budge, E.A.W., *Egyptian Language* (New York: Dover, 1971).

"Genetics Gold Rush," *The Economist*, June 13–19, 1981.

Montet, P., *Eternal Egypt* (New York: New American Library, 1964).

Office of Technology Assessment, US Congress, *Impacts of Applied Genetics: Micro-Organisms, Plants, and Animals* (Washington, DC: US Government Printing Office, 1981).

Pottle, F.A., ed., *Boswell's London Journal: 1762–1763* (New York: McGraw Hill, 1950).

Sackler, H., *Semmelweiss*, a play directed by Edwin Sherin, Washington, DC, October 1978.

Wallis, C., "Tampering with Beans and Genes," *Time*, October 19, 1981.

Walsh, J., "Biotechnology Book Reaches Agriculture," *Science*, November 27, 1981.

6

The Industrial Impact of the Biological Revolution

J. Leslie Glick

It is commonplace today to think of the 1980s as the decade of industrial biology. In fact, microbiology has been applied for millennia in the making of wine, beer, and bread. In the first half of the 20th century the chemical industry manufactured enormous quantities of ethanol, butanol, and acetone by means of fermentation. After World War II such fermentations were supplanted by lower cost chemical syntheses. As the cost of energy and of petrochemical feedstocks continues upward, the return to fermentation appears inevitable for production of commodity chemicals, given the trend of process improvements resulting from genetic and biochemical engineering.

During the 1960s and 1970s, the application of what is now regarded as classical microbial genetics (utilizing conventional mutation and selection techniques) enabled a diversity of products to be manufactured economically by means of large scale fermentation. Those products included amino acids, vitamins, enzymes, antibiotics, and steroid hormones. By 1980, bulk rates of those products represented an annual volume of several billion dollars worldwide. Over a billion dollars of sales were attributed just to two amino acids—mono-sodium glutamate (MSG) and lysine. MSG is used as a flavoring agent, and lysine as an animal feed additive to corn.

The new biological industrial revolution really represents a mix of old and new technologies. Its foundations are strongly rooted in the fermentation industry, but the discovery of recombinant DNA over ten years ago has resulted in a wide range of commercial opportunities that otherwise could not have existed. Molecular biologists now possess the tools to insert DNA from the genes of one organism into the DNA of another in such a precise way as to fashion a microorganism possessing the desired features of both. Such capability, in turn, creates new possibilities for production of commercial products by fermentation. Of course, advancements in industrial biology have not been limited to recombinant DNA technology. New developments are rapidly occurring in areas such as plant tissue culture and monoclonal antibodies. Nevertheless, I am focusing on recombinant DNA technology since it is that specialty of genetic engineering which, I believe, will impact most on the economy.

To make a recombinant DNA molecule, one first isolates DNA from two sources. One source is a microorganism, from which a circular, viral-like DNA, called plasmid DNA, is isolated. The second source is either another microorganism or an animal or plant cell. One then cuts both DNAs with the same enzyme. Alternatively, the second DNA may be chemically synthesized. The DNAs are then brought together and enzymatically joined. The resulting recombinant molecule is inserted back into the host microorganism in the form of a plasmid. The plasmid then replicates the newly formed recombinant DNA as if it were a normal part of the host genome and permits expression of the foreign gene's product.

There are two main rationales for applying recombinant DNA technology. One is to genetically modify microorganisms that are already capable of forming commercially desirable products to make them more efficient producers. Improvement of product yields results in lower costs of manufacture. A second rationale for applying recombinant DNA technology is to design microorganisms to produce products they would not naturally form. Recombinant DNA technology can thus be applied to the development of a wide range of product lines in diverse industrial settings. It is likely to impact heavily on the pharmaceutical, chemical, energy, agricultural, food processing, mineral extraction, and waste treatment industries. Product lines affected would include amino acids, vitamins, enzymes, steroids, peptide hormones, antibiotics, vaccines, petrochemical intermediates, fuels, and pesticides.

Of course, recombinant DNA technology alone will not produce the above products. Fermentation technology is what is required. Batch fermentation is commonly employed, with capacity sometimes exceeding 100,000 gallons per fermentor. More recently, continuous flow bioreactors, loaded with immobilized microbial cells or enzymes, have been designed to convert specific feedstocks or raw materials to desired end products. Such bioreactors are relatively small columns, in contrast to the large fermentation tanks.

For large-scale production by means of batch fermentation, cultures of the final production microorganism are grown in stages, from small volume seed cultures through a series of incubations. The culture volume increases by a factor of 20 during each incubation stage. The fermentation medium might typically consist of corn syrup, autolysed yeast extract, ammonia and salts. Most of the ingredients are premixed and pumped into the fermentor. After addition of any supplemental ingredients and adjustment of the pH, the final stage seed inoculum is also pumped into the fermentor. A typical fermentation batch has a duration of less than one week, with a minimum of about 2-1/2 days plus six hours allowed for turnaround of the fermentor. One fermentor could thus serve for a maximum of 128 batches per year, even allowing for down time due to maintenance and repairs. During the first quarter of the fermentation, rapid culture growth conditions are maintained. When a high cell titer is achieved, culture conditions are adjusted to promote a high level of production of the desired product.

It is interesting to compare the culture development required to produce equivalent amounts of product per year by batch fermentors and by continuous flow bioreactors. Use of the bioreactors will lower the costs for cell development by reducing the number of times the microorganism must be grown up. In one example applied at Genex, we have estimated that to produce over five million pounds of a particular specialty chemical in batch would require 50 fermentations per year, if each batch were as large as 100,000 liters. Alternatively, use of an immobilized enzyme procedure would require only one fermentor run of 6,500 liters to load a dozen 20-liter bioreactors with enough enzyme to last between one to two years prior to replacement with fresh enzyme. The 240-liter bioreactor volume would be sufficient to produce, by means of a continuous process, five million pounds of the desired specialty chemical within a year. Such a difference between batch and continuous processes clearly points out the advantages of reduced capital costs, less pollution, and lower energy requirements of the bioreactors.

Process Economics

The following examples illustrate how the utilization of genetic and biochemical engineering techniques may offer economic alternatives to current modes of manufacture of high value, intermediate value, and even low value products. The data behind these examples are described in detail elsewhere.[1]

Proteins are generally very expensive to manufacture by means of current processes, so they are clearly the most obvious targets for applying rival processes. Such proteins include enzymes, e.g., calf rennin, for making cheese; blood factors, e.g., human serum albumin, for treating patients with burns or in shock; hormones, e.g., human and bovine growth hormones, for treating patients with hypopituitary dwarfism and for increasing feed efficiency and milk production in dairy cows, respectively; and newer therapeutics, e.g., interferons, for treating patients with viral diseases and certain types of cancer. Current costs of manufacture may range from $1,000 per pound for rennin extracted from the lining of calf stomachs to $20 billion per pound for alpha interferon purified from human white blood cells. For all of these proteins the conventional production processes do not involve fermentation at all. Because their production is dependent on availability of the appropriate human or animal tissues, supplies are limited. The genetic engineering of microorganisms to produce such proteins will therefore result in much greater quantities produced and in considerably lower costs of manufacture.

Ideally, the genetically engineered microorganism should be capable of exporting the desired protein into the fermentation medium, as revealed by an appraisal of the respective process economics of intracellular production versus that of protein export. Recovery efficiencies are much higher for extracellular than for intracellular production, e.g., 80% versus 50%. Labor costs are strikingly lower in the case of extracellular production, largely due to the higher recovery efficiencies.

In addition, the production per unit volume of fermentor mash is related to cell density in intracellular production. To grow cells to high density requires a lot of raw materials, which is obviously inefficient in terms of conversion of raw materials to the desired protein. By comparison, an efficient protein-exporting cell is significantly more productive in yielding the specific product. This results in a second glaring contrast between intracellular and extracellular production of proteins, namely, the cost of raw materials. Primarily because of the differences in the costs of raw materials and labor, the unit costs for producing a typical protein intracellularly (at a level of 5% of the cell's total protein) and extracellularly (at a concentration of 10 grams per liter) amount to $580 per pound and $124 per pound, respectively, in a plant capable of producing 25,000 pounds per year. The cost of manufacturing an animal or human protein by means of microbial fermentation may thus be reduced over 40% if produced intracellularly and around 90% if produced extracellularly.

Cost estimates for antibiotic production by conventional and genetically engineered strains illustrate the potential for commercialization of recombinant DNA technology with respect to intermediate value products. Antibiotic production costs decline with increasing productivity of the organism, resulting in an increase in the concentration of the antibiotic in the fermentation broth. A problem that exists with antibiotic production by conventional fermentation techniques is that a large proportion of the raw materials employed in the fermentation is consumed by the organism for growth. Only a small proportion of the raw materials is utilized for the production of the secondary metabolite, the antibiotic. Since raw materials constitute a major portion of the expense of producing antibiotics, more money is spent on nutrients supporting growth than on the organic nutrient that is channeled to antibiotic production. This suggests that a wide margin exists that could be exploited for elaboration of the antibiotic; that is, diversion of raw materials from support of the microorganism to production of the antibiotic may be a possible route to improved antibiotic production.

We have estimated that in a plant producing 5.9 million pounds of an antibiotic per year, a conventional fermentation yielding a concentration of 1.2% antibiotic in the fermentation beer results in a unit cost of $13.32 per pound. By means of recombinant DNA techniques, increased production gives a final concentration of 6.0% antibiotic in the beer with a unit cost of $6.14 per pound. This decreased unit cost is the direct result of the increased efficiency of the genetically engineered organism. Expenditures for raw materials are thus cut by 50%. Hence, capital investment for equipment, buildings and storage is significantly lower; the fermentors are much smaller; and utility costs for the equipment drop sharply.

Importance in Low Value Products

A final example is offered, demonstrating the importance of biochemical engineering in the process economics of relatively low value products, such as amino acids. If a microbial strain is suitably engineered genetically, one can then choose to develop an immobilized cell system. As explained above, utilization of an immobilized cell system rather than batch fermentation may enable costs to drop significantly. Costs in the cell development area of batch fermentation are capital intensive, but adoption of the immobilized cell system can result in significant cuts in expenditures for equipment and buildings. Thus, for a two million pound plant producing a typical amino acid, the unit cost for batch fermentation totals $4.15 per pound, in contrast to the unit cost of $2.85 per pound for the immobilized cell system.

Two amino acids, methionine and lysine, are sold in bulk as animal feed additives. Prior to 1960, the selling prices of methionine and lysine (in 1980 dollars) were around $8 and $24, respectively. Twenty to 30 years later, those selling prices have gradually dropped to around $1 and $2, respectively. Increased production efficiencies sharply reduced the cost of manufacture, thereby enabling these amino acids to be sold at prices which spurred demand. The annual amounts sold worldwide increased 70-fold for methionine from 1954 to 1980 and 450-fold for lysine from 1958 to 1980. While the improvements in manufacturing economies predated any contributions from recombinant DNA technology, the examples are worth noting because recombinant DNA technology has recently been applied to lowering the costs of manufacturing tryptophan and threonine, two other important amino acids. Tryptophan and threonine have been sold at prices as high as $40 to $50 per pound but will not be used extensively as feed additives to corn until the prices drop around 90%.

The following simplified equations illustrate how the price of tryptophan is determined for its use in a chicken feed based on corn meal. The corn meal is supplemented with methionine, lysine, and tryptophan at such proportions to yield a feed with nutritional value equivalent to that of nonsupplemented soybean meal. If the price of the complete corn meal formulation is lower than the price of soybean meal, then the corn meal formulation will represent a more attractive feed to the farmer. In this example, the nutritional value of 119 pounds of corn meal supplemented with 1.06 pounds of methionine, 5.06 pounds of lysine, and one pound of tryptophan is equivalent to that of 167 pounds of soybean meal.

$$167\ S = 199\ C + 1.06\ M + 5.06\ L + T$$

$$T = 167\ S - 119\ C - 1.06\ M - 5.06\ L$$

where S, C, M, L, T = price ($ per lb) of soybean meal,
corn meal, methionine, lysine, and tryptophan, respectively.

It is clear from the above equations, that the price of tryptophan is highly sensitive to fluctuations in the price of soybean meal and corn meal. For every penny per pound increase in the price of soybean meal and every penny per pound decrease in the price of corn meal, the price of tryptophan could rise $1.67 and $1.19 per lb, respectively. In addition, the price of tryptophan is sensitive to the continuing reduction in the cost of lysine (due to steady improvements in lysine production). Thus, for soybean meal, corn meal, methionine, and lysine priced at $0.12, $0.06, $1, and $2 per pound, respectively, tryptophan would have to be priced at or under $1.72 per pound in order to be sold as a feed supplement. However, as the cost of lysine approaches that of methionine, i.e., $1 per pound, tryptophan could be priced as high as $6.78 per pound.

Near Term Prospects

A number of accomplishments of recombinant DNA technology already are well on the way toward commercialization. I have previously reviewed opportunities that could impact on the pharmaceutical,[2] agricultural,[3] food processing,[4] chemical,[5] and energy[6] industries. Pharmaceutical and agricultural (particularly animal health care) applications are nearest towards commercialization. Table 1 lists some examples that pertain to the pharmaceutical and agricultural industries. These are briefly described below, the pharmaceutical applications being discussed first.

Human insulin is one of the most widely publicized products resulting from recombinant DNA technology. Eli Lilly has now begun to market human insulin in the UK and the US. Microbially derived human insulin, which should gradually replace the conventional bovine and porcine insulins, will be readily available independently of the supply of animals or the upward trend in the population of diabetics. Human insulin may also have a therapeutic advantage. Administration of animal insulin induces an allergic reaction in perhaps 5% of insulin-treated patients.

There is a pressing need for another microbially produced hormone, human growth hormone. Most individuals suffering from hypopituitary dwarfism cannot afford to purchase the extraordinarily expensive and scarce hormone currently available. Its supply is limited since it is obtained from pituitaries of human cadavers. Genentech is testing microbially derived human growth hormone in clinical trials pertaining to hypopituitary dwarfism. Production of human growth hormone by means of fermentation will result in large enough supplies to test its efficacy in novel therapeutic situations, such as an inducer of wound healing.

Recombinant DNA technology has also been applied to the development of microbially produced interferons. Alpha, beta, and gamma interferons have been produced in three different types of microorganisms — *Escherichia coli, Bacillus subtilis,* and *Saccharomyces cerevesiae* — by Biogen, Cetus, Genentech and Genex. Biogen, in particular, has produced fermentation batches on the order of 30,000 liters each. These anti-viral agents are now being tested not only against various troublesome viruses but also for potential efficacy as anti-cancer agents. Their production by fermentation will result in copious quantities at drastically reduced costs. As explained above, alpha interferon, for instance, could be produced under $1,000 per pound by means of fermentation instead of the outrageous cost of $20 billion per pound using conventional technology.

Another area well under attack by recombinant DNA technology involves the production of human vaccines which either are not readily produced or are relatively ineffective when produced by means of conventional technology. Hepatitis B virus antigens, for example, have already been expressed in microorganisms, and Biogen has licensed a microbially produced hepatitis vaccine to Green Cross in Japan.

With respect to antibiotics, more efficient yields may result from genetic manipulation of those microbes which produce a particular enzyme that is capable of modifying either a known antibiotic or an inactive precursor to a more effective form. Recombinant DNA technology has thus been used to enhance the level of

TABLE 1. Some Accomplishments of Recombinant DNA Technology in Various Stages Toward Commercialization by the Pharmaceutical and Agricultural Industries

Product	Use
Pharmaceutical Industry	
Human insulin	Control of diabetes
Human growth hormone	Control of dwarfism
Human interferons	Anti-viral and anti-cancer agents
Hepatitis B vaccine	Prevention of hepatitis
Penicillin G acylase	Manufacture of penicillin
Agricultural Industry	
Bovine growth hormone	Enhanced milk production
Scours vaccine	Prevention of diarrhea in newborn calves and pigs
Hoof and mouth disease vaccine	Prevention of hoof and mouth disease
Single cell protein	Animal feed
Tryptophan	Feed additive
Threonine	Feed additive

penicillin G acylase in *Escherichia coli.* This enzyme catalyzes the conversion of benzylpenicillin to 6-aminopenicillanic acid.

The agriculture-related products listed in Table 1 all pertain to animal health care. Animal vaccines represent the first such group of products on which recombinant DNA technology has begun to impact. Intervet International, a division of the Dutch Company Akzo, already markets a vaccine using antigens produced by genetically engineered *Escherichia coli* to prevent scours, a diarrhea in newborn pigs and calves. A similar product developed by Cetus has just begun to be marketed by Nordeen Laboratories, a subsidiary of SmithKline Beckman. Vaccines produced by genetically engineered microorganisms to prevent hoof and mouth disease are currently being developed separately by Genentech, in collaboration with the US Department of Agriculture-Agricultural Research Service, and by Biogen. Hoof and mouth disease vaccine should be on the market within a couple of years.

A variety of animal growth hormones produced in genetically engineered microorganisms are currently under development by several companies. Genentech and Genex, for example, have separately designed microorganisms to produce bovine growth hormone. As mentioned above, bovine growth hormone is known to stimulate milk production in dairy cows but is too expensive to use if obtained from its conventional source. Genex has also worked with microorganisms to produce porcine and ovine growth hormones. The former hormone is expected to be used commercially to increase meat production in pigs. The latter hormone would be used to enhance wool production in sheep.

For more than 15 years, technology has been developing for using microorganisms as an inexpensive source of protein for supplementing animal as well as human diets. Such protein, known as single cell protein, represents a potential commodity to the feed and food markets. It will not be widely utilized, however, until its cost becomes more competitive with that of traditional sources of protein. A variety of raw materials have been evaluated for conversion to single cell protein, ranging from petrochemicals to waste effluents from various industrial processes. Imperial Chemical Industries has applied recombinant DNA technology to improve the conversion of methanol to single cell protein in one bacterial species — *Methylophilus methylotrophus*. Imperial Chemical Industries is now the major producer of commercially available single cell protein.

The remaining two products listed in Table 1 represent two amino acids, tryptophan and threonine. The rationale for producing these two compounds for use as feed additives has been explained above. Genex has applied recombinant DNA technology to produce both these products in higher yields than would otherwise be possible without such genetic engineering.

Due to the steady stream of advances in genetic and biochemical engineering technologies, industry is increasingly looking to biotechnology as the source of a broad spectrum of product lines. Table 2 offers one example of an industrial strategy to provide a mix of products by means of biotechnology. This particular strategy, which has been pursued at Genex, focuses on the development of fine chemicals for application to food processing, feed additives, animal health care, and waste and water treatment. Table 2 lists 13 products in various stages toward commercialization, which are distributed among four product lines. We have estimated that the world market for the 13 products amounts to around $600

TABLE 2. Example of an Industrial Strategy to Commercialize a Mix of Products by Means of Biotechnology

Product Line	Product	Use	World Market Value* ($ Millions)	
			Present	1990
Amino acids	Aspartic acid	Sweetener raw material	9	60
	Phenylalanine	Sweetener raw material	12	240
	Threonine	Animal feed additive	16	32
	Tryptophan	Animal feed additive	10	121
Vitamins	Vitamin B_{12}	Feed/food additive	20	40
	Vitamin C	Feed/food additive	410	800
Enzymes	Calf rennin	Cheese manufacture	80	130
	Glucose isomerase	Sweetener manufacture	40	60
	Lipase	Water treatment	0	40
	Peroxidase	Waste treatment	0	20
Peptide hormones	Bovine growth hormone	Milk production	0	70
	Ovine growth hormone	Wool production	0	35
	Porcine growth hormone	Meat production	0	55
			597	1703

*In 1980 dollars

million today and may reach $1.7 billion by 1990. Such growth is projected for the following reasons. Some of the products currently lack a market due to prohibitive costs of manufacture. The ability to produce these economically will spur demand. Likewise, the demand for most of those products which currently are marketed is still highly sensitive to further reductions in manufacturing costs. Finally, new applications have been found for some of the products, thereby leading to new markets.

Four amino acids are listed in Table 2. The rationale for producing two of them, tryptophan and threonine, has been explained earlier. The other two—aspartic acid and phenylalanine—serve as raw materials for the artificial sweetener aspartame. Only recently permitted by the Food and Drug Administration for sale in the US, aspartame is expected to capture the major share of the market for artificial sweeteners. As a result, the demand for aspartic acid and phenylalanine will increase commensurately. Phenylalanine is relatively expensive, compared to aspartic acid. Those companies that can improve the manufacture of phenylalanine in order to lower its cost will attain a competitive edge in the marketplace.

The two vitamins represented in Table 2, vitamins B_{12} and C, are currently marketed for human and animal applications. The market for vitamin C is already quite large. However, an efficient microbial production process for vitamin C that uses inexpensive chemical raw materials might result in effective price competition with current, combined microbial-chemical synthetic processes. In contrast to the market for vitamin C, that for vitamin B_{12} is relatively small. The price of vitamin B_{12}, unlike that of vitamin C, is very expensive, as much as $3,000 per pound. Clearly, manufacturing improvements which lead to lower selling prices would impact positively on market share and would also sharply stimulate demand.

Table 2 lists four enzymes. The largest single market for these is that for calf rennin. Used to coagulate milk in the manufacture of cheese, calf rennin costs around $1,000 per pound. Moreover, its supply has not kept pace with demand, since veal consumption has decreased concomitantly with an increase in cheese consumption. Naturally occurring microbial substitutes have been introduced in the market, but these are not used for higher quality cheeses because they generate undesirable flavors. Calf rennin produced in microorganisms, however, should offer both quality and price advantages.

Another enzyme of major commercial value is glucose isomerase, a catalyst used in the manufacture of high fructose corn syrup. The increasing acceptance of high fructose corn syrup in the soft drink industry has stimulated demand for glucose isomerase. Opportunities exist for incorporating various potential improvements into this enzyme and the processes in which it is used.

The other two enzymes listed in Table 2 currently lack a market because their applications are still being developed. One enzyme, lipase, may be used to remove fatty materials from water, sewers, and water treatment facilities. The other enzyme, peroxidase, may be used to eliminate undesirable colors from industrial effluents.

Finally, Table 2 lists three animal growth hormones. As explained above, all three hormones are too expensive to market if obtained from their conventional sources—pituitary glands. Microbial production will remove that economic hurdle.

Long Term Outlook

In the future, recombinant DNA technology will impact significantly on areas in which molecular genetics was previously thought to be inapplicable, e.g., the genetic engineering of plants for desirable qualities such as nitrogen fixation, insect and disease resistance, and higher yields. Such research and development programs are already being pursued on plants by commercial organizations.

Certainly gene therapy for human hereditary disorders represents one of the most ambitious applications of recombinant DNA technology. Considerable progress has already been made in cell culture and animal models. Sickle cell anemia and Cooley's anemia are the hereditary disorders most likely to be treatable by purified gene preparations. As many as 7,000,000 persons worldwide may suffer from these two disorders alone, both of which are frequently fatal at an early age.

Other long term targets of recombinant DNA technology include conventional commodity chemicals such as ethanol and ethylene oxide, as well as what may develop into novel commodity chemicals such as pullulan and poly-beta-hydroxybutyrate. These latter two compounds, which are naturally occurring, microbially produced polymers, represent nature's answer to plastics. They may readily substitute for existing products that are synthesized chemically from ethylene. Pullulan, for example, can be used in a variety of applications, such as a food wrap, packaging film, textile yarn, gauze, adhesive, or even a molded product. Moreover, it is both biodegradable and nontoxic.

Ultimately recombinant DNA technology will be applied in an even more imaginative context.[7] Our technology will gradually evolve from mimicking nature, i.e., producing naturally occurring gene products but in foreign host microorganisms, to devising truly novel gene products that conceivably never existed before in nature. We now have the capability to construct genes for virtually any protein sequence we desire. This capability will exist in increasingly powerful form in the next few years. We shall progress to designing custom-made proteins and enzymes. We would like to carry out chemical reactions biologically for which no enzymes are known to exist at present. The limitation to this approach at present is not so much our ability to manipulate genes as it is our knowledge of what specific alterations in the primary structure of an enzyme molecule lead to what specific changes in the tertiary structure of that enzyme. The ability to define a chemical reaction, design the necessary catalytic scheme, and then specify the three-dimensional folding of a specific amino acid sequence for the enzyme will mark our true mastery of biochemistry. It seems very likely that genetic engineering will be employed to alter enzymes so as to give them more desirable properties within the next ten years, and employed to design genes coding for enzymes with entirely new properties within the next 20 years. Biotechnology will then have outgrown its infancy.

With respect to forecasting the economic impact of recombinant DNA technology on society, I will simply summarize some projections that have resulted from a comprehensive technology assessment program developed at Genex.[8] In 1980, we evaluated some 500 different products. We speculated that the application of recombinant DNA technology by industry worldwide would result in annual sales of those products amounting to $40 billion in the year 2000. The corporate investment required to develop this business was projected to be $24 billion by the year 2000. Corporate income in the year 2000 was projected at $3.5 billion for all 500 products. Such earnings would lead to an estimated 15% return on investment in the year 2000 ($3.5 billion return divided by $24 billion investment).

We then attempted to quantitate how the public might benefit on the sales of those 500 products. The financial return to the public in the year 2000 on those sales of products whose manufacture might be affected by recombinant DNA technology was projected to be $4.7 billion. This figure combined cost savings (to the consumer) on currently existing product lines, amounting to $2.4 billion, and increased tax revenues due both to sales of new products as well as market growth of existing products, amounting to $2.3 billion. We assumed that the investment to be made by the public to permit such a financial return would be in the form of government funded basic research programs. Presumably by 1990 the findings resulting from this public research investment would provide the underpinnings for the subsequent commercialization envisaged for the year 2000. We estimated that cumulative government funds spent in the 1970s and those projected for the 1980s on recombinant DNA research would reach $3 billion worldwide. Thus, the projected public return on investment pertaining to the application of recombinant DNA technology might approach 160% in the year 2000 ($4.7 billion return divided by $3.0 billion investment). It is striking, therefore, that the public could benefit extraordinarily well relative to the private sector, realizing an overall tenfold higher percentage return on an investment one eighth that of the private sector.

References

1. J. Leslie Glick, "The Cost Reducing Effects of New Medical Technologies: Health Care in the Age of Recombinant DNA Technology and Applied Genetics" in *Proceedings of the International Conference on New Technologies for the Reduction of Health Costs* (Sydney, Australia: Institute of Health Economics and Technology Assessment, in press, 1983). See also J. Leslie Glick, "Genetic Engineering and the Chemical Industry" in *Investing in Biotechnology* (Beds, England: Oyez Scientific & Technical Services Ltd., 1982), pp. 65–82.
2. Glick, *op. cit.*, The Cost Reducing Effects of New Medical Technologies: Health Care in the Age of Recombinant DNA Technology and Applied Genetics."
3. J. Leslie Glick, M. Virginia Peirce, David M. Anderson, Charles A. Vaslet and Hung Yu Hsiao, "Utilization of Genetically Engineered Microorganisms for the Manufacture of Agricultural Products" in *Proceedings of the Beltsville Symposium VII, Genetic Engineering: Applications to Agriculture* (Washington, DC: US Department of Agriculture, 1983, in press).
4. J. Leslie Glick, "Impact of Recombinant DNA Technology on the Economy" in H.H. Fudenberg, ed., *Biomedical Institutions, Biomedical Funding, and Public Policy,* (New York: Plenum, 1983, in press).
5. *Idem.* See also Glick, *op. cit.,* "Genetic Engineering and the Chemical Industry."
6. J. Leslie Glick, "Fundamentals of Genetic Technology and Their Impact on Energy Systems" in R.F. Hill, ed., *Proceedings of the Energy Technology IX Conference — Energy Efficiency in the Eighties* (Rockville, Maryland: Government Institutes, Inc., 1982), pp. 1470–1474.
7. J. Leslie Glick, "Benefits Accuring From Genetic Engineering" in *Science and Society in the 80's: Genetic Engineering and Its Impact* (Starkville, MS: Mississippi State University, 1982), pp. 1–3.
8. Glick, *op. cit.,* "Impact of Recombinant DNA Technology on the Economy."

Part III

7

Introduction
Joseph G. Perpich

This section presents a series of articles which focuses on the wide-ranging potential of biotechnology. Having opened the volume with perspectives on research and development programs now present in universities, government agencies, and the industry itself, the authors in this section will discuss the impact of biotechnology on the workings of government: Congress, the courts, and the executive branch.

Harrison Schmitt

The first contributor is former Senator Harrison Schmitt, a former astronaut and key formulator of science policy, who currently lives in New Mexico and is lecturing and consulting at the present time. Senator Schmitt holds an undergraduate degree from the California Institute of Technology and a doctorate in geology from Harvard University. He joined the NASA Apollo program in 1965, serving in 1972 as the lunar module pilot for Apollo 17. The only geologist in the space program, he was appropriately the last man to step on the moon during the Apollo program.

In 1976 he was elected Senator from New Mexico and became the ranking minority member of the Subcommittee on Science, Technology and Space (Committee on Commerce, Science and Transportation). During the 95th and 96th Congresses (1977–81) he worked closely with Committee Chairman Adlai Stevenson in the numerous oversight hearings involving recombinant DNA research policy under the National Institutes of Health Recombinant DNA Research Guidelines. During the 97th Congress, Senator Schmitt rose to chairman of that subcommittee and also became chairman of the Appropriations Committee's Labor, Health and Human Services and Education Committee. Thus, he has had both legislative oversight on science policy as well as responsibility for appropriations for the National Institutes of Health.

Senator Schmitt's article for this volume presents issues for debate based on his scientific training and his role in senatorial science policy. He notes the recombinant DNA public policy debates, especially how they engaged Congress in the late 1970s, and the resulting questions, among them, what impact might governmental regulations and new priorities have on scientific and technological innovation? He also looks at issues for determining the appropriate roles for government and the private sector. One central question that Senator Schmitt poses concerns the government's capacity to assess risk and manage it, and he reviews, among others, his bill which proposed to set up demonstration projects in risk analysis in the federal government to allow better definition of the basis for federal regulations. He concludes with a discussion of other legislative areas of priority to science and technology, specifically how biotechnology might spur innovation and productivity in the United States.

David L. Bazelon

David L. Bazelon, the second contributor, is Senior Circuit Judge of the United States Court of Appeals for the District of Columbia (retired). He served as Chief Judge of the Court from 1962 to 1978. A graduate of Northwestern University Law School, Judge Bazelon was an Assistant Attorney General in the Justice Department during 1947–49, and was named by President Harry S. Truman to the United States Court of Appeals in 1949 — then the youngest appointee to the federal bench. In 1978, when he stepped down after 16 years as Chief Judge, *The Washington Post* noted that "his contributions to American law in the twenty-eight years he had served on that Court have been remarkable." His influence on criminal, civil and regulatory law has been broad, especially so for the latter, as the US Court of Appeals for the District of Columbia over the past decade became increasingly involved in the review of federal agency regulatory decisions concerning health and safety.

It is from these perspectives that Judge Bazelon looks at the emerging biotechnology industry. He notes that this technology, as all others, must be judged in terms of the risks and benefits — but recognizes that such assessments are necessarily affected by our values based on long-standing assumptions, priorities and dispositions. Judge Bazelon explores the differing "regulatory" approaches, including — in the absence of legislative or regulatory action — that of the common law. He notes the problems with the common-law approach in dealing with risks to health and the environment, providing his perspectives on the appropriate role of the courts through judicial review of legislation and attendant executive branch regulations. Finally, Judge Bazelon examines the role experts play in public policy — both in the regulatory forum and in the courts. He recommends action appropriate for biotechnology experts that might better assist sound public policy.

Joseph G. Perpich

The last article in this section is based on a speech I presented at the Washington-Baltimore Harvard Medical Bicentennial Day celebration in Washington, DC (1982). Beginning with an outline of the development of the NIH Recombinant DNA Research Guidelines and related federal policies, I next focus on federal activities today in biotechnology research. Looking also at the NIH planning strategies, which were based on a desire to use governmental research resources most effectively to maintain the momentum in the biological revolution, I discuss models for government, university, and industry initiatives and interactions. This review of governmental programs highlights the enormous federal commitment to biotechnology research and development, a spread that covers virtually every federal agency with an R&D component. Finally presented are some models that might enhance this investment for government, universities and industry — at all levels.

Thus, from the three perspectives representing the branches of our government, these three articles, which appeared in *Technology in Society* in 1983 (Vol. 5, No. 1), highlight the impact of biotechnology on these institutions and, necessarily, on all governmental institutions, whether they be international, federal, state or local. Further, these articles set the stage for the next section, which will be devoted to the relationship of biotechnology to patent, corporate, and regulatory law, and the evolving legal relationships among the partners in this research.

8

Biotechnology and the Lawmakers

Harrison Schmitt

The almost total lack of perspective in the Congress of the United States on scientific matters and the perils that await scientists who dabble in politics have rarely been more obvious than during the first major public policy debate related to recombinant DNA (rDNA) in 1977 and 1978. From a historical perspective, this was one of the most important and certainly most instructive episodes in the interaction of science and public policy.

In 1971 Nobel Laureate James Watson chose the topic of cloning and genetic engineering for his address to the House Science and Astronautics Committee (now the House Science and Technology Committee).[1] This was, perhaps, the first inkling that the members of Congress had that a revolution in biology was beginning — a revolution in which Congress would become deeply involved in the years to come.

The recombinant DNA technique was "invented" in the laboratory in 1973 by two scientists in California, Herbert Boyer and Stanley Cohen.[2] Within a year of this "invention," a report prepared by the Congressional Research Service noted that concerns were already surfacing regarding possible hazards with the new technique, even through recombinant-like processes were known to occur in nature.[3]

The hypothetical, or alleged, risks surrounding research with rDNA were brought to the public's attention in mid-1973 by a specific group of research scientists who were themselves involved in such research. Led by Doctors Maxine Singer of the NIH and Dieter Söll of Yale, this group called on the National Academy of Sciences to consider the risks surrounding this new research.[4]

The National Academy of Science subsequently convened an expert panel, chaired by Nobel Laureate Paul Berg, whose final recommendations included:

- a moratorium on some rDNA experiments until the risk could be adequately assessed;

- the establishment of an Advisory Committee on recombinant DNA within the NIH; and

- the convening of an international conference of experts to consider the question of risk and recombinant DNA.[5]

One criticism of this early panel that I had was that it included no epidemiologists, no experts in infectious disease, who could assess the real risk of human (or animal) disease from the possible creation of inadvertent recombinant DNA pathogens. I suggested later that a geologist or geophysicist might even have been included to evaluate any effects on the evolutionary process.

In any case, the international conference was held in 1975 in Asilomar, California (another conference was held at the same location in 1973 on the hazards of working with animal viruses; this conference went virtually unnoticed outside the scientific community). The conference report outlined physical and biological containment standards and recommended that these standards be applied to experiments based on their *perceived hypothetical* risk.[6] This report was formalized into the first set of guidelines for recombinant DNA research by the newly created NIH Recombinant DNA Advisory Committee.[7]

An Issue of Public Policy

What made recombinant DNA an issue of public policy in the first place? It was the possible or alleged risk of the technique, not the potential benefits. This was reflected in much of the legislation that was later introduced in the US Congress and in the attitude of many members of the public, notably several environmental groups, local officials in Cambridge, Berkeley, and other cities, and the media.

The Congress did not play a major role in the early stages of the recombinant DNA issue. The House Science and Technology Committee directed the preparation of a series of reports on the subject, and the Senate Committee on Labor and Public Welfare (now Labor and Human Resources) held hearings in 1975, following the Asilomar Conference, and again in 1976. However, in 1977 Congress decided to take a direct hand in matters. Early that year Bills were introduced in both the House and the Senate that were rather moderate. However, as time progressed and the legislative process worked its will, the Bills became increasingly stringent. The final versions in the Senate (S. 1217, sponsor: Kennedy) and in the House (H.R. 7897, sponsors: Rogers and Staggers) were actually longer and more detailed than the Food, Drug and Cosmetic Act.

In November 1977 hearings were held on the regulation of recombinant DNA research by the Subcommittee on Science, Technology and Space, chaired at that time by Senator Adlai Stevenson, and on which I served as ranking minority member.[8] Throughout the 95th Congress, many bills were introduced in both the Senate and the House dealing with this issue. Other committees, notably the Senate Committee on Labor and Human Resources and two House Committees (Science and Technology and Interstate and Foreign Commerce) had also held hearings and even marked up and reported legislation to their respective Houses. However, at the time of our hearings in November, we seemed further away from a legislative answer than ever before. And this was in a political atmosphere when most people believed an Act would be on President Carter's desk before mid-summer 1977.

This extraordinary level of interest by Congress was due to a number of reasons:

- the importance and complexity of the issue — we were faced with decisions on revolutionary research in biology, the scientific basis for which virtually no one in either the House or Senate and few in the media or general public understood; to this we must add the political aspirations of some of the elected officials who were the principals in the debate — both at the national and state and, particularly, the local levels;

- the vigor of the lobbying effort by the scientific community — this episode probably represents the "rites of passage" for the scientific community into the realm of public policy; for the first time, the total freedom of scientific inquiry was questioned — some say "threatened" — and scientists streamed into the halls of Congress to testify and educate; and

- the implications of this issue for science policy in general and the governance of basic research — the federal government is responsible for funding a large amount of all scientific research, particularly state-of-the-art research, some of which might be classified as theoretically hazardous; thus, the federal government's spending policies in themselves create a form of *de facto* national science research policy. Thus, the potential power of the government to limit or inhibit the development of new knowledge by controlling the purse strings is vast.

The Doubtful Necessity of Legislation

During our hearings in 1977 and later during 1978, when the report on our hearings was being prepared and bills still proliferating, I remained unconvinced that any legislation was necessary in light of two pieces of scientific information:

- The risk of inadvertent pathogenesis by crippled *E. coli* strains appeared extremely remote. In the summer of 1977 a workshop on risk assessment in recombinant DNA experimentation was held at Falmouth, Massachusetts. This meeting included internationally recognized experts in clinical infectious disease, epidemiology, virology, and genetics. An important consensus was arrived at unanimously by the assembled group, *i.e.*, there was *no* theoretical or experimental evidence that the strain of *E. coli* used for rDNA experiments (K-12) could become pathogenic, transfer dangerous plasmids to other host bacteria, or colonize the human intestine or any other ecological niche.[9]

● Recombinant or recombinant-like processes had been shown to occur naturally among bacteria and the advantage of rDNA technology was only in speeding up the otherwise natural processes of plasmid transfer and recombination. The acquisition of natural antibiotic resistance by bacteria is a good example. Work by several population geneticists at the time also indicated that, at least in *E. coli*, the carriage of a piece of DNA (plasmid) exerted negative, or unfavorable, natural selection against the carrier. From the beginning it was evident that bacteria use their own restriction enzymes to destroy foreign DNA (specifically bacterial viruses) — a sort of primitive "immune system" — while protecting their own DNA. It is these enzymes that make the entire technology possible — and fast.

During the latter part of 1977 (after our hearings) and throughout 1978, I worked successfully to delay any legislation until these new scientific data could be evaluated and the NIH could revise its Recombinant DNA Guidelines to reflect the current scientific thinking. Interestingly, the original Guidelines contained *no* provision for any sort of periodic revision of the document. Thus, when the NIH held hearings on revision of the Guidelines in 1978, the resulting version contained not only a mechanism for periodic modification of the guidelines, but also a provision enlarging the Recombinant DNA Advisory Committee by adding six non-biologists.[10]

We know now, looking back, that this whole recombinant DNA controversy illustrated three broader problems: the lack of general understanding of public policy issues with a large technical component; timely theoretical and experimental assessment of risk; and the determination of proper government action, if any.

This short analysis of the attempts to legislate limits on recombinant DNA research can assist us in answering questions that still have great relevance for science policymakers today:

● *What impact do government regulation and, perhaps more importantly, funding priorities have on scientific and technological innovation and development; and, as a corollary, if scientific and technological development are stymied, what benefits to the society are ultimately lost?*

  — As we have seen, the lack of government funding for high-risk (either in terms of payoff or hazard) research is tantamount to prohibition of such research. Moreover, while appropriate regulation may be desirable in some areas (toxic substances, nuclear energy, production of foods and drugs), excessive regulation stifles basic research and limits public benefits from it.

  I was extremely disturbed at the suggestion, made in 1977, that the provisions of section 361 of the Public Health Service Act (PHSA, 42 U.S.C. 264) dealing with communicable diseases be applied to the regulation of recombinant DNA research. Not only was there no scientific basis for such an application, but the possible penalties for violation were extraordinary for purely hypothetical and unproven risks. The use of 361 would certainly have created an undesirable precedent for broad intrusions into the research environment and thwarted any effort at innovative research and development in this area.[11]

Surely, when a society stifles creative basic research, the innovative capacity of that society is diminished and thus the society itself is weakened. If we had ultimately regulated recombinant DNA to death (and many advocated a complete prohibition until *all* risks had been removed) our society today would be without genetically engineered human insulin, human growth hormone, a vaccine for neonatal calf scours (a life-threatening disease that ultimately reduces our food supply), and other potentially life-enhancing products.

● *If it becomes necessary to regulate basic research, what are the appropriate roles for the scientific and business communities, the public, the Congress and the regulators?*

—In fact we don't know the answer to this question. All in all, the interaction between these various groups concerning recombinant DNA was certainly beneficial with regard to mutual understanding, but, since no real regulations were ever promulgated (only the Guidelines), we cannot really evaluate the future roles of the above players. However, the modifications of the Guidelines that have occurred since 1976 did elicit public, scientific, industrial, Congressional and regulatory comment and, because of this continuing interaction, the system is perceived to be "working."

● *What are the capabilities of the government to actually assess risk and subsequently manage it?*

—Up to the present, the risk assessment programs of the federal government have been fragmented. Different regulatory agencies have different methods of quantifying risk (varying in comprehensiveness and quality) and different ways of making regulatory or policy decisions based on their quantification. There is little governmentwide coordination and virtually no effort has been made to first evaluate the various methods that agencies use and then select the best way to assess, quantify, and analyze risk. I worked to remedy this situation (along with Congressman Don Ritter of Pennsylvania) through legislation (S. 3006) which proposes to set up demonstration projects in risk analysis in the federal government, to evaluate these projects, and coordinate and disseminate information on risk analysis to interested parties. Perhaps if we had had a way to accurately assess the perceived risk of recombinant DNA research in those early days, we could have avoided some of the misunderstanding and acrimonious debate that subsequently took place, but then, hindsight is a wonderful gift.

● *Should different guidelines (or regulations) be applied to industrial and academic laboratories?*

—Initially this was a cause for some concern to the Congress, not only because the original NIH recombinant DNA guidelines applied only to

NIH-funded research, but because state, and particularly local, governments were moving to regulate, often in a rather draconian fashion, both academic *and* industrial research. It was hoped that some federal legislation would preempt what in 1977-78 was fast becoming a patchwork of differing state and local laws.

At the time, I certainly was not convinced of the wisdom of pre-emptive federal legislation, particularly as it applied to the private sector, because of the question of encroachment on the rights of state and local governments to protect the health and safety of citizens within their political jurisdictions from particular harm. Today this same question, but in a different context — that of use or prohibition of certain pesticides — will probably be answered by the Supreme Court (*NACA vs. Rominger*, 500 F. Supp. 465 (E.D. Calif., 1980)). Thus, in that light, we were wise in 1977 not to pre-empt the States.

So far, however, there has been no operational difference in the application of DNA "regulations," *i.e.*, the recombinant DNA Guidelines, to either industrial or academic research. All of the new companies revolving around this technology, as well as the larger firms, have voluntarily and unanimously complied with the NIH Guidelines. We have not yet had to face the question of industrial abuse or outright and flagrant violation of the NIH recombinant DNA Guidelines.

It is interesting to note that industrial manufacturing processes for food, drugs, cosmetics and toxic substances have been subject to great regulation, while university-based research in these same areas has been virtually unregulated. Perhaps because of the markedly different goals of industry and academia, this approach is the correct one. If so, the present situation in biotechnology is an aberration. If, on the other hand, both academic and industrial R&D are merely different means to the same end, they should be subject to the same sorts of regulation (or absence of regulation). It is a thorny question and one with no easy answer in the absence of any systematic and continuing study in this area of public policy.

● *What are the consequences of regulation on the technological activities of other nations and on the US competitive economic position relative to them?*

Early in the history of recombinant DNA almost all nations conducting significant levels of research were already doing, or considering, these rDNA experiments. Several, notably Britain, gave serious consideration to this issue. With the exception of Britain, however, all adopted a "wait until the NIH acts" attitude. Britain produced a risk-assessment study (the Ashby Report) even before the Asilomar conference.[12] The British government then set up the Genetic Manipulation Advisory Group (GMAG), which adopted rules very similar to the NIH Guidelines. Countries have generally adopted either NIH or GMAG-type guidelines and, in some cases, a "recombinant" form of the two. Interestingly the outcome in the US of no special legislation regulating this research has been paralleled in Europe.

Ironically, our major competitor, Japan, has not relaxed its guidelines to the extent we have in the US, and the Japanese recently complained of being at a "competitive disadvantage" with regard to their ability to conduct this research. However, in general, the US and its major trading partners have remained virtually on a par with regard to real or *de facto* regulation of recombinant DNA research. Clearly the US has the lead in most areas of biological research and technology (one possible exception being Japan in fermentation), and the private sector is certainly striving to maintain our lead.

Perhaps this entire episode in science policymaking demonstrated most clearly—particularly to the scientific community—that if we wish to maximize beneficial scientific research and yet preserve both the welfare and the confidence of the public in that research, we must better define and utilize our community of experts and science policymakers to ascertain the facts, inform political opinion-makers of these facts, and ensure that the media and the public understand the judgments and policy decisions based on those facts.

There is always some lag between technological developments and societal adjustment to them, "future shock," as it were. Thus, public participation in scientific issues requires an informed public, including policymakers, in order to be meaningful and constructive. Scientists were perhaps a little taken aback, much like the sorcerer's apprentice, to discover that the debate they started amongst themselves soon acquired a life of its own, out of their control. Today the scientific community is still learning how to interact with policymakers and how important these interactions are. There is, unfortunately, a long way to go.

In the past two or three years, Congress and the courts (*Diamond vs. Chakrabarty, In re Bergy,* etc.) have adopted a much more benevolent, even encouraging, attitude toward the development of biotechnology—which now encompasses not only recombinant DNA technology, but other biological technologies, such as monoclonal antibodies and immobilized enzymes. I would like to outline some of the principal issues that may directly or indirectly impact on both the development of the still-fledgling biotechnology industry as well as on university-based research in this area:

- Government Patent Policy: In the 97th Congress, I introduced, along with Representative Allen Ertel of Pennsylvania, the Uniform Science and Technology Research and Development Utilization Act (S. 1657), or "The Patent Bill." This legislation would have extended to all contractors regardless of size the first right of refusal to title to inventions developed with federal funds. Currently this first right of refusal only accrues to universities, small businesses, and non-profit organizations (P.L. 96-517), which account for only a small percentage of total government-funded R&D. The end result of such legislation would have been to increase commercialization of products and processes that have languished on federal agency shelves without this incentive for investment. Unfortunately, both the House and Senate versions of the bill died at the end of the 97th Congress.

- Risk Assessment: As we have seen, the perception of risk (its quantification and analysis) played a key role in the recombinant DNA controversy. There is a clear need for reproducible, accurate and continuing risk assessment methodologies, particularly in the agencies making regulatory decisions based on these assessments. This need will become more important, not less so, as the public demands safe food, drugs, energy sources—anything that touches their lives. The Risk Analysis Demonstration and Research Act would have set up a coordinated research and demonstration program for risk analysis under the aegis of the Office of Science and Technology Policy in the Executive Office of the President. Such a program could have led to improved approaches to federal regulation, particularly in the fields of health and the environment, and this certainly would have included regulatory decisions on the products of biotechnology.

- Science, Math, and Engineering Education: The US today faces two critical problems, neither of which has an easy answer—first, the shortage of competent, well-trained teachers of pre-college science and mathematics (we can't train enough, nor can we retain the ones we've trained), and second (undoubtedly related), the alarming extent of scientific illiteracy among our young people. There are a variety of approaches that have been suggested to solve these problems. The National Science and Technology Improvement Act of 1982 (S. 2809) focused on three critical areas for science and mathematics: instrumentation, research awards to young university faculty members, and support for pre-college teachers to improve their instructional skills and subject knowledge. In writing the Bill, we required that at least one-half of the funding in the areas of research awards and pre-college teacher improvement be from non-federal sources. Hopefully this would have served to leverage the federal dollars to the greatest extent without making unreasonable fund-raising demands on the recipients. Although the bill itself failed to complete the process of becoming law, several of the programs it contained were incorporated into the FY 1984 program plan of the National Science Foundation.

Recent Acts of Congress which may have major positive effects are the Small Business Innovation Development Act (P.L. 97-219) and the Economic Recovery Tax Act (ERTA) of 1981 (P.L. 97-34). They both provide for a variety of economic incentives for industrial R&D.

So far the federal government has encouraged biotechnology by a program of enlightened participation and, to some extent, anticipation.

Endless Opportunities

What lies ahead? The view from here seems endless in terms of opportunities available for applications of biotechnologies. Already the first genetically engineered products (human insulin, human-growth hormones, and calf scours vaccine) have reached or will reach the marketplace in the next few months. In recent years, the application of genetic engineering techniques to agriculturally important crops has yielded large quantities of important information and has revolutionized our thinking about plants and plant genetics.

Potentially plant genetic engineering could be the new recombinant DNA revolution of the 1980s and 1990s. In addition, new micro-organisms have been developed via recombinant techniques that will "eat" non-nuclear hazardous waste, although their use is still in the research mode. Under laboratory conditions, the ability to degrade 2,4,5-T and 2,4-D has been successfully transferred to the common soil bacterium, *Pseudomonas*. One can imagine the time when chemical spills and hazardous waste dumps will be cleaned up by specially engineered "bugs."

Other uses that are being investigated are for mineral leaching from low-grade ore, biomass conversion and ethanol production (as alternate energy sources), commodity chemicals and single-cell proteins (hailed as a potential long-term solution to the world food shortage).

Of course, the work in the health area will continue, with a major effort directed at the application of recombinant DNA technology to humans. Recently a Presidential Commission released a report dealing with the ethical considerations that revolve around human applications of "genetic engineering."[13] A major conclusion of this report, which was the subject of three days of Congressional oversight hearings, was that, while such research could lead to important breakthroughs in the treatment or even the cure of disabling genetic diseases and thus should proceed, the command for "full speed ahead" should be tempered with a careful consideration of the ethical questions that reflect public concern.

A Spur to Innovation

In addition, the biotechnology industry—like other "high-tech" industries, may represent an important spur to US innovation and productivity, qualities that have been singularly lacking in recent years. I have sponsored legislation (P.L. 97-367) that has convened a White House Conference on Productivity during this year. The conference is chaired by William Simon and has involved leaders from all segments of US industry, labor, academia, and all levels of government, including biotechnology and other "high-tech" industries, such as microelectronics and computer sciences.

At the present moment the role that Congress may play appears to be one of informed oversight, aimed at part at the education of members and staff. One issue that Congress will surely have to monitor closely is that of "deliberate release" into the environment of recombinant DNA organisms, a practice that until April 1982 was proscribed entirely by the NIH Guidelines. The Guidelines now, under specific procedures and review, permit "deliberate release."

The changes in the Guidelines reflect current scientific knowledge and the necessity, imposed by the need to test chemical-consuming "bugs" outside the laboratory. Thus, the new provision of the NIH Guidelines is a first step in permitting such advances. Congress will and must monitor the advancing research closely, determining over time whether legislation is needed.

Eventually the Congress may have to face legislatively the questions left unanswered by the Supreme Court decision on patentability in *Diamond vs. Chakrabarty*.[14] In the near term, initiatives in risk assessment, science education and patent reform must continue.

Certainly the interest in this subject among investors, legislators, scientists and the public remains high and the development of both the industrial applications as well as the basic research in all areas of biotechnology will be watched with great anticipation.

Acknowledgement

The author would like to acknowledge the assistance of Dr. Norine E. Noonan, American Chemical Society Congressional Science Fellow, in the preparation of this article. Dr. Noonan is currently on leave from the faculty of Georgetown University Schools of Medicine and Dentistry.

Notes

1. Watson, J.D., Proceedings Before the Committee on Science and Astronautics, US House of Representives, 93rd Congress, First Session, January 26, 27 and 28, 1971, pp. 336–366.
2. Cohen, S.N., A.C.Y. Chang, H. Boyer and R.B. Helling, Construction of Biologically Functional Bacterial Plasmids *in vitro, Proceedings of the National Academy of Sciences* 70: 3240–3244 (1973).
3. US Congress, House Committee on Science and Astronautics, Subcommittee on Science, Research and Development, *Genetic Engineering: Evolution of a Technological Issue*, 93rd Congress, Second Session, Supplemental Report 1 (1974), 215 pp.
4. Singer, Maxine, and Dieter Söll, "Guidelines for DNA Hybrid Molecules," *Science* 181 (1973), p. 1114.
5. Berg, Paul, *et al.*, "Potential Biohazards of Recombinant DNA Molecules," *Science* 185 (1974), p. 303.
6. Berg, Paul, D. Baltimore, S. Brenner, R.O. Roblin III and M.F. Singer, "Asilomar Conference on Recombinant DNA Molecules," *Science* 188 (1975), p. 991.
7. National Institutes of Health, Department of Health, Education and Welfare, "Recombinant DNA Research Guidelines," *Federal Register*, July 7, 1976, pp. 37902–37943.
8. US Congress, Senate Committee on Commerce, Science and Transportation, Subcommittee on Science, Technology and Space, Hearings on the Regulation of Recombinant DNA Research, 95th Congress, First Session, November 2, 8 and 10, 1977, Serial 95-52.
9. Gorbach, Sherwood, ed., "Risk Assessment of Recombinant DNA Experimentation with *Escherichia coli* K-12," *Journal of Infectious Diseases* 137 (1978), pp. 611–714.
10. National Institutes of Health, Department of Health, Education and Welfare, "Revised Guidelines Regarding the Conduct of Research Involving Recombinant DNA: Final Action," *Federal Register*, December 22, 1978, pp. 60080–60105; 60108–60131.
11. Schmitt, H., "A Scientist-Senator on Recombinant DNA Research," *Science* 201 (1978), pp. 106–108.
12. Secretary of State for Education and Science, Report of the Working Party on the Experimental Manipulation of the Genetic Composition of Microorganisms (Lord Ashby, chairman), London: Her Majesty's Stationery Office, Cmnd. 5880 (1975), 23 pp.
13. President's Commission for the Study of Ethical Problems in Medicine and Biomedical and Behavioral Research, *Splicing Life: A Report on the Social & Ethical Issues of Genetic Engineering in Human Beings* (1982), 77 pp.
14. *Diamond vs. Chakrabarty*, 447 U.S. 303 (1980); *In re Bergy*, 596 F.2d 952, (Court of Customs and Patent Appeals), 1979.

9

Governing Technology:

Values, Choices, and Scientific Progress

David L. Bazelon

Freud once praised the wonders of modern technology for enabling him to speak with his children living hundreds of miles away. On second thought, he noted that — were it not for the damn modern railroad — his family would not be so far away in the first place.

As a federal judge for more than three decades, I have developed a similar ambivalence toward new technologies. In my private life I enjoy the tremendous benefits that technological progress has brought. On the bench, however, I frequently hear the pleas of persons for whom a particular form of progress represents an onerous burden.[1] That perspective constantly reminds me that technological progress has its costs, and those costs rarely fall equally on us all. The burdens of progress are allocated — explicitly or otherwise.

My observations of earlier technologies make me confident that biotechnology will be a similarly two-sided coin. The promise of biotechnology offers hope to the most destitute people on earth. Through its miracles we can conceive of winning the eternal struggle against hunger and disease.[2] At the same time, the application of biotechnologies carries with it risks that are difficult to define, much less to assess. Many uses of biotechnology will involve the release into the open environment of life forms not currently found in nature. The effects of such microorganisms on the surrounding environment cannot be predicted with certainty — and might be catastrophic. In the event of disastrous consequences, moreover, it is likely that many people will be harmed who had little to gain from the use of the technology in the first place.

In short, the development and use of biotechnology involves decisions about what risks are worth taking for whose benefit. Competing values, each of which is held dearly when considered alone, must be traded off against each other. Selecting these trade-offs is invariably difficult because it forces us to expose priorities which, for a variety of reasons — political or otherwise — we would prefer were left unexposed.[3] In such circumstances, it is tempting to structure the decision-making process in a way that hides the clash of values. Methods of choice can be used to create an appearance either that we have avoided making a decision or that it was easy because the proper choice was clear. Both of these approaches are suspect, and the first is pure illusion.

When faced with a promising, but risky, technology, there is no such thing as a non-decision. Prohibiting the technology can preserve the status quo, but at a sometimes devastating cost to those who could have benefitted from it. Postponing a choice is sometimes less devastating, yet frequently entails other costs.[4] Both time and additional information can be expensive resources, and investing in them involves trade-offs as well.

Because difficult issues cannot be avoided, decision-makers frequently camouflage their choices as value-neutral ones. The desired appearance is often achieved by delegating decisions to institutions that convey an appearance of objectivity. A common example is the reliance on the workings of free market forces. But the determinations of the marketplace reflect its values, which *a priori* are no more neutral than any others.[5] I am reminded of the story of the elephant who shouted, "Everyone for himself!" as he stood in front of a pile of grain among a flock of chickens.

The appearance of value neutrality is also often created through the quantification of competing values to enable comparison of them in an "objective" manner.[6] It is a marvelous trick — simply paint the apple orange, then pick the fairer fruit — but an illusion at best. At bottom, all the difficult decisions about biotechnology will rest on value-laden assumptions, priorities, and predispositions. Shall we release into the ocean a bacteria that cleans up oil spills? What effect will it have on fish? Who gives a damn about fish? I prefer white beaches — though I do have a tender spot for salmon. Uncertain risks coupled with unpleasant trade-offs — but decisions need to be made. Who shall make them, and how?

The Decision-Makers

Many scientists feel that the regulation of science should properly be left up to the individual scientist. As the publisher of *Scientific American* once wrote, "A scientist can accept no authority but his own judgment and conscience. . . ."[7] If outside interference is to exist at all, according to this view, it should come exclusively from within the scientific community. For only those peers understand the complex issues sufficiently to assess them rationally.

That view predominated until relatively recently. Scientific and technological progress was seen as inevitable and inherently desirable, and we marvelled at the fantastic rate of technological advance. Since World War II, however, government, science and technology have become increasingly interdependent. The reasons for this symbiosis are several. First, the costs of projects, such as constructing an atomic bomb and exploring the universe, are so enormous that only the government has the resources to foot the bill. With funding inevitably comes some supervision. Second, we have become increasingly conscious of the adverse effects that frequently accompany technological developments. Decisions about whether to beat those costs inevitably involve value choices. Such choices certainly fall beyond the exclusive domain of scientists and engineers. Although their expertise is essential for assessing the costs and benefits of particular innovations, it provides no special qualification for determining the appropriate balance between the two.

Scientists frequently object to outside scrutiny, and their complaints are not frivolous. It costs a lot to pay the salaries of the bureaucrats who look over the scientists' shoulders. Moreover, such surveillance can, of course, impede or even stifle research. Many scientists would prefer to return to the prior "unregulated" research environment. They assert that the government should avoid imposing health and safety measures on society, or better still, get out of the risk-control business altogether.[8]

Such suggestions frequently confuse the health-risk problems with the agencies and regulations created to deal with them. Society has always regulated health risks — and always will. Long before regulatory agencies, our system of private lawsuits served to control risk-taking — to encourage some activities and discourage others. In a sense, one cannot really "deregulate" these activities. One can only redistribute their benefits and burdens.[9] Biotechnology can be promoted, for example, by relaxing research regulations and shielding scientists from liability arising out of its use. Likewise, biotechnology can be discouraged by extending the reach of private lawsuits in an attempt to redress indirect harms caused by its use. When an unnatural life form is released into the environment, for example, who shall be liable for harms it may cause — the scientist who developed it, the company that produced it, the customer who released it, or all of them? And what sorts of harms should they be liable for?

Regulation in Common Law

If agencies and legislatures fail to address the risks of biotechnology — either to deter them or compensate their victims — the courts will likely do so, perhaps even more vigorously, in deciding private lawsuits. I can easily imagine the effects of biotechnology producing expansive doctrines in the common law of nuisance and tort. There are many modern examples of the common law evolving to enable private lawsuits to address the harms of new technologies. Tort actions for cancer, for example, have encountered many obstacles because of difficulties in establishing causation, and the fact that cancer frequently develops many years after exposure to carcinogens, when much of the evidence needed to establish liability has been lost. In responding to this problem, the California Supreme Court recently permitted a cancer victim to sue a group of manufacturers of the hormone DES, even though she did not know which one had produced the specific drug that had injured her.[10]

Regulation through the common law has many drawbacks. It has a substantial impact on science, technology, and the economy generally. It "regulates" and constrains just as surely as an agency does. It can cause researchers to follow a variety of unnecessary practices, simply to avoid lawsuits. It also imposes extremely high costs in damages, insurance, and attorneys' fees. Moreover, judicial regulation cannot provide the consistency, rationality, or political responsiveness offered by a consciously designed and clearly articulated legislative solution. A courtroom is not the place to decide such complex and controversial issues of fact and policy. And judges are not the appropriate persons to decide them.

The problem is not just that these scientific issues are complicated; courts have long grappled with complicated issues in reviewing actions by the FCC, SEC, ICC, CAB, and scores of other governmental regulatory agencies. These more traditional administrative matters, however, involve issues with which all judges have at least a speaking familiarity. Increasingly, the caseload of our courts involves challenges to federal administrative action relating to the frontiers of technology. Expanding health and safety regulations and increasing citizens' suits have drawn the courts into such difficult questions as: What level of exposure to known carcinogens is safe for industrial workers?[11] Shall we ban the Concorde SST,[12] DDT,[13] or lead in gasoline?[14] How shall society manage radioactive wastes?[15] I dare say that most judges do not have the knowledge and training to assess the merits of competing scientific arguments involved in these issues, and that it is hardly a task for on-the-job training.

More important, regulation through common law doctrines places decisions concerning appropriate trade-offs among competing values in the hands of judges. As an independent branch of government lacking in political accountability, the judiciary should feel reluctant to play this role. I know that I do. I realize that such a role is, to a certain extent, inevitable in a common law system such as ours. But the evolution of the common law through judicial innovation creates fewer problems of legitimacy when that evolution occurs in small increments over a long period of time. Such evolution is more likely to reflect a consensus of values and to permit legislative intervention when it frustrates the majoritarian will.

In recent years, however, the rapid advancement of powerful technologies has created new legal problems on a scope and at a rate that overwhelms the ability of the common law to respond in a coherent, legitimate manner. The nature of the technological risks at stake cannot be addressed sensibly on a case-by-case basis. Many harms, particularly those to the environment, take years to manifest and often cannot be traced to their sources. Those damages must be addressed through regulation to prevent harm rather than litigation to redress it. Control of such risks cannot be left to the *ad hoc* value choices of judges through possibly inconsistent determinations at trial.

The temptation to place difficult value choices on judges is particularly great concerning ethical questions which, because they seem less complex, seem more within the competence of judges. Advances in the biological sciences have already presented several such questions and are sure to generate many more. Questions concerning the legal definition of death, a patient's right to die, a patient's right to demand treatment such as *in vitro* fertilization, all present difficult moral questions in a technological context. Biotechnology presents similar ethical questions arising out of, for example, the foreseeable potential to alter the genetic makeup of man.[16] Although perhaps few persons would object to tampering with genes to eliminate tragic birth defects, who should define a birth defect? If we have the capability to make everyone six feet tall, should we do so? Should our tampering with the "natural order" deprive someone who would otherwise be of unusual intellect of his privileged position?

Ethical Versus Technical Issues

Some of these issues present questions of constitutional law, which the judiciary must decide. For the most part, however, judges are not the appropriate persons to decide such issues. In the context of the courtroom, moreover, ethical questions tend to be avoided in favor of the technical issues involved. In the recent controversy over the patentability of living microorganisms,[17] for example, the legal arguments have focused on the intended coverage of patent laws and the distinction between an invention and a living organism.[18] Little debate has considered the moral question whether to extend the concept of proprietary rights to commercial use of new life forms.

Government regulation will not necessarily eliminate the problems I have mentioned concerning regulation through private lawsuit. But it may obviate the courts' need to confront many intractable risk problems and ethical dilemmas. If Congress has consciously made an ethical choice concerning a particular technology, if it has weighed the scientific evidence, assessed the economic effects, and tested the political winds concerning a particular risk, *or* if it has delegated these tasks to a regulatory agency, specifying the procedures, policies, and standards to be applied, then courts will more readily defer to those legislative decisions.

Although the appropriate role for the courts in the regulation of biotechnology is limited, it is nevertheless important. It consists principally of judicial oversight of actions by administrative agencies. In playing that role, however, courts and judges must recognize the implications of their institutional strengths and limitations. It makes no sense to rely upon the courts to evaluate the scientific and technological determinations of agencies. There is perhaps even less reason for the courts to substitute their own value preferences for those of the agency, to which the legislature has presumably delegated the decisional power and responsibility. The limits of judicial competence in technical fields and the judiciary's lack of political accountability preclude these roles.

The contribution that courts can make in this process is in monitoring the decision-making processes of agencies to make sure that they are open, thorough, and rational.[19] In this role, courts rely on their institutional strengths. The independence of the judiciary, which delegitimizes judicial choices among competing values in the first instance, is an advantage in the role of monitor. Likewise, lawyers are very familiar with the rights of interested parties to be heard, requirements of notice, openness, and disclosure. These concepts form the core of the administrative process. The legal training of the bench and bar—although wholly inadequate for evaluating scientific evidence—is good preparation for overseeing that process.

The Need for Public Debate

The goal of judicial oversight is to facilitate peer review and legislative and public oversight. By forcing an agency to articulate the factual basis and rationale of its decisions, they can be evaluated by other experts in academe, government, and industry. Individuals and groups that differ with the agency's value choices can make their views known in the various public forums. Such public debate gives reason to hope that erroneous decisions can be changed in light of new information or changing preferences.

The need for public ventilation of difficult value choices reminds me of an extraordinarily painful experience that I had many years ago at the University of Washington School of Medicine. I had been invited to sit in as an anguished group of doctors, confronted by the mind-boggling cost and limited availability of life-saving renal dialysis, sentenced some of their patients to death so that others could live.[20] A pair of massive oaken doors hung tightly closed in the background. After the doctors had made their horrifying choices, one anxiously asked me if I had any suggestions. Feeling impotent in the face of these awesome moral dilemmas, I could offer only one thought. "See those doors over there," I said. "Keep them open. Let the public know how you have made these decisions. Show the public your assumptions, your uncertainty, and your value choices. Let them share your burden."

When I have told this story at other times, many people have pointed out that when horribly difficult choices must be made, there may be value in not knowing how it is done. Should a person sentenced to die because of a shortage of renal dialysis machines have to face the added trauma of knowing why he was not chosen to live? Must he be told that it was because he was old, or poor, or uneducated, and considered less "valuable" than other candidates for treatment?

Such an argument is only persuasive to me if one assumes that certain decisions are fixed, when, in fact, they are not. One must assume, for example, that the number of available renal dialysis machines is set in concrete. In fact, that number reflects an allocation of society's resources away from other uses. Can we really assume that, if the public understood the impact of their decision and the choices it implies, they would not make a different choice? Might not the public decide to build one less nuclear warhead in order to save a certain number of individuals who happen to have defective kidneys? Unless we know that the answer to that question is negative, we must not encourage the public to remain uninformed.

Having argued for full disclosure of values and uncertainty, I realize that institutional pressures often militate against this approach. Uncertainty is messy. It detracts from the simplicity of presentation, ease of understanding, and uniformity of application. To focus on uncertainty is to invite paralysis. To disclose it is to risk public misunderstanding or opposition. Such unattractive possibilities encourage policymakers to ignore uncertainties and compensate for them by, for example, making intentionally inflated estimates of risk. They might incorporate extremely conservative assumptions about the shape of a dose-response curve for low levels of a harmful agent. But such tactics do not erase the uncertainty inherent in many decisions, and an added safety margin may tilt the balance away from the best alternative. The goal is accurate forecasting to enable a comparison among alternatives. Where scientific estimates are highly tentative and filled with uncertainty, those uncertainties must be fully disclosed and considered as part of the package.

Consideration of Unknowns

I do not mean to suggest that public oversight should impede agency action in the face of uncertainty. For some activities, the magnitude of potential harm and the probability of its occurrence may be essentially unknown. That is certainly the case concerning much research in biotechnology and the application of it. Many risk estimates depend upon future contingencies of human behavior or other highly complex and unpredictable variables. Historical experience may be totally lacking, as it was when biotechnology first began. Even the best risk estimates are subject to an unknown degree of residual uncertainty and may thus overstate or understate the dangers involved. And many times an agency must act in circumstances that make a crap game look as certain as death and taxes. In such situations an agency need only disclose the uncertainties that it faces and explain why action is necessary in spite of such risk. As long as the agency satisfies the level of certainty required by Congress, the courts should not interfere.

When scientists participate in the public debate over biotechnology — as I hope that they will — they must keep in mind the specific role that they play. They are not, unless so designated, the policymakers. Their role is not to make conclusions concerning the appropriate trade-offs among risks, but rather to make clearer what the estimated trade-offs are. What the public needs most from any expert, biologists included, is his wealth of intermediate observations and conceptual insights adequately explained. Decision on the ultimate questions must be left to the public decision-making process.

My experience with the regulatory system suggests that scientists are uncomfortable in this role. Scientists who appear in the public arena all too often focus on little more than making conclusory pronouncements. Either they omit any real discussion of underlying observations and methods of inference — or they drown such discussion in a sea of jargon. To paraphrase Lewis Carroll, they use "labels as shrouds rather than guides." They tell us a particular innovation is safe, rather than *how* safe and *why*. They ignore the basic fact that a conclusion that a technology is "safe" reflects a host of value choices about the relative importance of such diverse concerns as the health of a particular industry or company, the severity of the problem addressed by the technology, and the value of the things that the technology might harm. Just as a doctor's decision to send a patient home from the hosptial might be influenced by overcrowded conditions in the hospital, a biologist's recommendation that a particular microorganism can be safely released into the environment may be affected by considerations independent of the risk that the release might cause harm.

In short, conclusory statements are of little use in making ultimate decisions that must be left to the public arena. Policy questions are multi-dimensional. They involve scientific, moral, and social judgments. Conclusory statements cannot be digested by the decision-making process. Simply put, they fail to provide the facts that the public needs to mix with its moral and social judgments. Scientists must recognize the right of the public to make basic value and risk choices.

Making Decisions Openly

As an interested observer of the biotechnology controversy, the development of the NIH Recombinant DNA Research Guidelines strikes me as an optimistic example of how decisions on controversial issues can be made openly.[21] At the time that work on the Guidelines began, the situation in biotechnology had all the ingredients for developing into an impassioned political stalemate. From the scientists' perspective, the principle of free scientific inquiry was at stake over a technology that may be the most powerful tool ever devised for biological advance. From the public's perspective, biotechnology involved the release of new life forms into the environment with potentially catastrophic consequences—a fact that many of the researchers acknowledged in 1973 when they adopted a self-imposed moratorium on certain kinds of research. These conflicting interests created strong pressure to gloss over the potential risks and convince the public not to be interested.

Instead, the public was invited to participate in the development of guidelines for regulating the research. At the opening of a public hearing on the Guidelines, Dr. Frederickson, Director of the NIH, observed, "Recombinant DNA research brings to the fore problems of public scrutiny of the process and the progress of basic science. . . . Procedural safeguards with a full exploration of relevant facts and possible alternatives must be the hallmark of the scientific process, if we are to retain the trust and the whole-hearted support of society."[22] In addition to public participation, the NIH developed an environmental impact assessment in compliance with the National Environmental Policy Act of 1969.[23] The assessment gave thorough consideration to the environmental effects of research to be conducted under the Guidelines.

In retrospect, it seems that the initial Guidelines and the concern that produced them were over-reactions. The dangers appear less catastrophic, and controlling them less difficult than at first imagined. But the process has shown itself very flexible in meeting changed perceptions of risk. The Guidelines have been continuously reconsidered, and many have been revised or removed.[24] Public hearings concerning possible changes have been held at every stage. This process continues apace.

Critics of public participation will point to the delays that it causes to suggest that public involvement should be kept to the minimum. It is possible, for example, that the NIH Guidelines could have been written more expeditiously if they had been developed with greater secrecy. I suspect, however, that the later the public had been asked to participate, the greater their sense of suspicion would have been. The willingness of scientists to describe their uncertainties, to allow public participation and comment, has produced considerable good will toward biotechnology. Despite some early sensationalism, I think that the public has performed responsibly. The Guidelines seem to have produced an outcome that allows research in biotechnology to continue in a manner that scientists can live with and that the public finds worthwhile.

The regulatory experience of biotechnology contrasts sharply with the history of relations among the public, the government, and industry concerning nuclear power.[25] That history has been shaped in part by a continued insistence by industry and the regulators that the issues involved are beyond the grasp of the public. To prevent the public from creating obstacles to the new technology, the regulating agencies have continuously tried to restrict public participation. When mistakes have been made — as inevitably they have been — the public's disappointment is reinforced by their lack of involvement. This feeling has certainly hampered the ability of the NRC to provide the public with assurances that it has adequately considered the safety-risks of nuclear power. The fact that many mistakes of the past had been foreseen by public groups whose warnings were ignored has added fire to the public's cynicism. The resulting uneasy atmosphere has hampered the growth of both nuclear technology and the nuclear power industry.

The Requirement of Openness

It follows, therefore, that openness is in everyone's best interest. When issues are controversial, any decision will fail to satisfy large portions of the community. But those who are dissatisfied with a particular decision will be more likely to acquiesce in it if they perceive that their views and interests were given a fair hearing. If the decision-maker has frankly laid the competing considerations on the table — so that the public knows the worst and the best — he is unlikely to find himself accused of high-handedness, deceit, or cover-up. Scientists cannot afford, for the public will not tolerate, the handling of these vital matters in a manner that invites public cynicism and distrust.

In the final analysis, the requirement of openness and candor in controlling risky technologies reflects our society's democratic values. Power in the society resides with the people. The freedom enjoyed by scientists and industry to explore is given by the public and can be taken away. In this sense, the prerogatives of a technology depend upon the public good will. False reassurance, unjustified confidence, and hidden agendas will only encourage the public to exercise its ultimate veto power. Our people have always been prepared to accept risks and pursue the greater good of society. Progress can hardly be achieved any other way. It was Thomas Jefferson who once said, "If we think the people not enlightened enough to exercise their control with a wholesome discretion, the remedy is not to take it from them, but to inform their discretion."[26] Choices will be made despite uncertainty and despite their social disruptions and dislocations. To preserve the good will on which biotechnology depends, however, society must be informed about what is known, what is feared, what is hoped, and what is yet to be learned.

Notes

1. See *e.g.,* Keene v. Insurance Co. of North America, 667 F. 2d 1034 (D.C. Cir. 1981) (asbestos); American Federation of Labor v. Marshall, 617 F. 2d 636 (D.C. Cir. 1979), *aff'd in part, vacated in part sub nom;* American Textile Mfrs. Institute, Inc. v. Donovan, 452 U.S. 490 (1981) (cotton dust); Environmental Defense Fund, Inc., v. Ruckelshaus, 439 F. 2d 584 (D.C. Cir. 1971) (DDT).

2. See Krause, "Is the Biological Revolution a Match for the Trinity of Despair?", *Technology In Society* 4:4 (1982).

3. See generally Calabresi and Bobbitt, *Tragic Choices* (1978), discussing different methods societies use to make "tragic" choices.

4. See Environmental Defense Fund, Inc. v. Hardin, 428 F. 2d 1093 (D.C. Cir. 1970) (reviewing agency inaction rather than action).

5. Perhaps most significantly, market determinations reflect an existing distribution of wealth. See Calabresi and Bobbitt *supra* note 3, at 81-129 (discussing methods of modifying market allocation systems and the flaws of such modifications).

6. This approach is seen most clearly in attempts at cost-benefit analysis. The flaw with such analysis in many contexts is well known. It stems from the difficulty of identifying and quantifying many costs and benefits; the inevitably arbitrary nature of valuations of human life or health; the problem of interpersonal and intergenerational comparisons of utility; and many others. See P. Schuck, *Regulation: Asking the Right Questions,* 11 Nat'l J. 711 (1979): National Academy of Sciences, *Decision Making for Regulating Chemicals in the Environment* 39–44 (1975) (report prepared by the National Research Council); E. Quade, *Analysis for Public Decisions* 25–26 (1975).

7. Sinsheimer and Piel, *Inquiring Into Inquiry: Two Opposing Views,* Hastings Center Report, August 1976, at 19 (statement by Piel).

8. In 1976 the National Science Foundation asked directors of leading American research institutions for their views on the state of American science. A recurring response was an objection to excessive regulation of scientific activities, and to bureaucratic "meddling" in the scientific domain. See National Science Board, National Science Foundation, *Science at the Bicentennial: A Report from the Research Community* 63–69 (1976). In the words of one participant in the study, "the ever increasing bureaucracy . . . will in the not too distant future completely eradicate our Nation's world position in research and technology." Id. at 66 (response of the Director of Los Alamos Scientific Laboratory).

9. See L. Friedman, *A History of American Law* 409–27 (1973) (arguing that the common law of torts changed during the industrial revolution in order to protect emerging industries from some forms of liability); M. Horwitz, *The Transformation of American Law* (1977) (same); but see Schwartz, *Torts Law and the Economy in Nineteenth-Century America: A Reinterpretation* 90 Yale L.J. 1717 (1981) (arguing that nineteenth century tort law did not favor industry).

10. Sindell v. Abbott Laboratories, 607 P. 2d 924, *cert. denied sub nom.* E.R. Squibb & Sons, Inc. v. Sindell, 449 U.S. 912 (1980) (cancer victim permitted to sue six of 200 DES manufacturers because six defendants constituted 90% of the market).

11. Industrial Union Department, AFL-CIO v. American Petroleum Institute, 448 U.S. 607 (1980) (benzene).

12. See Environmental Defense Fund, Inc. v. Coleman, No. 76–1105 (D.C. Cir. May 19, 1976) (unreported).

13. See Environmental Defense Fund, Inc. v. Ruckelshaus, 439 F. 2d 584 (D.C. Cir. 1971).

14. See Ethyl Corp. v. EPA, 541 F. 2d 1 (D.C. Cir.) (en banc), *cert. denied,* 426 U.S. 941 (1976).

15. Natural Resources Defense Council, Inc. v. NRC, 547 F. 2d 633 (D.C. Cir. 1976), *rev'd sub nom.* Vermont Yankee Nuclear Power Corp. v. Natural Resources Defense Council, Inc., 435 U.S. 519 (1978).

16. See, *e.g.,* President's Commission for the Study of Ethical Problems in Medicine and Biomedical and Behavioral Research, *Splicing Life: A Report on the Social and Ethical Issues of Genetic Engineering with Human Beings,* ch. 3 (1980).

17. See Diamond v. Chakrabarty, 447 U.S. 303 (1980).

18. *Id.* at 309–10. The Court explicitly refused to consider the policy issues raised by the parties, saying that they should "be addressed to the political branches of the Government, the Congress and the Executive, and not to the courts." *Id.* at 317.

19. For several years my colleagues on the D.C. Circuit and I have engaged in a lively debate about the standards that should govern judicial review of administrative action in scientific areas. See, for example, the five separate opinions in Ethyl Corp. v. EPA, 541 F. 2d 1 (D.C. Cir.) (en banc), *cert. denied,* 426 U.S. 941 (1976), in which our court upheld regulations issued by the EPA Administrator requiring annual reductions in lead content of gasoline. In large part, the debate has focused on the extent to which judicial review should examine only agency procedures followed in taking an action or whether it should also review the substance of the action. Compare Bazelon, *Coping with Technology Through the Legal Process,* 62 Cornell L. Rev.

817 (1977) with Leventhal, *Environmental Decisionmaking and the Role of the Courts,* 122 U. Penn L. Rev. 509 (1974).

20. For additional discussion of this allocation system, known as the "Seattle God Committee," see Calabresi and Bobbitt, *supra* note 3, at 187–88.

21. For a more complete history of the development of the guidelines, see Perpich, "Industrial Involvement in the Development of NIH Recombinant DNA Research Guidelines and Related Federal Policies," 5 *Recombinant DNA Technical Bulletin* 59 (June 1982).

22. *Id.* at 60.

23. 42 U.S.C. § 4321 (1976).

24. See 43 *Fed. Reg.* 33042–33178 (July 28, 1978) (proposed revisions); 43 *Fed. Reg.* 60080–60105 (1978) (final revisions).

25. For a richly detailed account of the nuclear power controversy, see S. Tolchin and M. Tolchin, *Dismantling America: The Rush to Deregulate,* ch. 6 (Boston: Houghton Mifflin Company, in press).

26. Letter from Thomas Jefferson to W.C. Jarvis (September 28, 1820), reprinted in 7 *Writings of Thomas Jefferson* 177, 179 (H. Washington, ed., 1855).

10

Genetic Engineering and Related Biotechnologies:

Scientific Progress and Public Policy

Joseph G. Perpich

In 1974 I finished my training in psychiatry and the law, joining NIH as Associate Director for Program Planning and Evaluation in 1976. My first assignment was to direct the staff effort to develop the NIH Recombinant DNA Research Guidelines. As it turned out, I was grateful for both my legal and psychiatric training; I needed every bit of each. For I quickly discovered that the greatest of all needs was "due process" — a lawyerly specialty if ever there were one! And my years spent as a concerned listener, standing apart from the fray, served me well as the monitor and recorder of debates that frequently took an emotional turn.

Those debates in my tenure at the NIH concerned at first the development of the NIH Recombinant DNA Research Guidelines, but later they extended to many related federal policies. Using that as background, I will focus here on federal activities today in biotechnology research, for current research strategies and the exploding areas of biotechnology are not immune to fiscal constraints, which have prompted some reexamination of the role government, universities, and industry should play in this research. The "genetic engineers" are largely drawn from the academic molecular biology departments, but they are becoming established businesspeople in their own right — whether as key personnel in a biotechnology company or as prime members of teams in the larger industries that have a

Joseph G. Perpich, M.D., J.D., guest editor of the current Technology In Society *series, "Biotechnology — Impact on Societal Institutions," is Vice President for Government Affairs of Genex Corporation, Rockville, Maryland. This article is based on a speech by Dr. Perpich at the Washington-Baltimore Harvard Medical Bicentennial Day Celebration, Cannon Building, United States Capitol, November 13, 1982.*

biotechnology component. My focus, therefore, is to merge my NIH experience with the new perspectives I see now in industry itself.

Procedural Development of NIH's Recombinant DNA Research Guidelines

Let me remind you of the circumstances in which Dr. Fredrickson, the NIH Director, and the NIH mounted the effort to develop what are now NIH's Recombinant DNA Research Guidelines. Two great value systems were on a collision course: on the one hand were concerned scientists seeking to uphold an embattled, but sacred, principle — free scientific inquiry. In opposition stood concerned environmentalists from many backgrounds, riding — in the mid-1970s — a surge of public support for protecting the environment. The questions became: how much scientific freedom, and at how much risk to the public? Thus was the scene set for due process. Beyond — or in addition to — this "clash" of cultures was an NIH imperative of its own: to avoid an overt regulatory role.

Our basic approach for handling the many matters toward our goal of Recombinant DNA Guidelines was one, that in retrospect, we feel was successful. It became, in fact, a model as we wrestled with subsequent research policy issues at the federal level. The process had three important phases: (1) assembling a "kitchen cabinet" of senior NIH staff, who advised the NIH Director on primary objectives and pointed out all of the ensuing inter-related issues we could think of; (2) airing the reviews of those issues in an open forum representative of wide scientific and public opinion (for the most part, that forum was the Advisory Committee to the NIH Director and the Federal Interagency Advisory Committee on Recombinant DNA Research); and (3) creating the public record of the proceedings themselves with related documents to support the conclusions reached. In the end we had a set of procedures we were comfortable with for policy formulation. It addressed the diverse aspects and tensions involved, set standards, and monitored basic research in a rapidly developing field. But the NIH Guidelines stopped short of external regulation.

Our efforts to develop the Guidelines were prompted by a 1975 meeting of scientists at Asilomar. Those in attendance requested the NIH promulgate guidelines to ensure protection of the public health and environment from unforseen dangers of recombinant DNA research: micro-organisms with newly inserted genes. Some were apprehensive about the potential risk from this research to those within the laboratory or, if the micro-organisms were set free intentionally or accidentally, to the public and the environment. By June 23, 1976, the most difficult steps were complete: the NIH Guidelines for Recombinant DNA Research were issued (published the following July 7 in the *Federal Register*).[1] Dr. Fredrickson noted at the time that "the issue was how to strike a reasonable balance — in fact a proper policy 'bias' — between concerns to 'go slow' and those to progress rapidly."

But publication of the Guidelines did not quiet all the related issues. Three major ones were to require significant work over the next three years: (1) the potential need for legislation to regulate all recombinant DNA research; (2) the issuance of an environmental impact statement on the Guidelines themselves; and (3) the potential patenting of recombinant DNA research inventions. (For a detailed history and elaborated set of remarks on the development and revision of the NIH guidelines for recombinant DNA research, see J.G. Perpich, "Industrial Involvement in the Development of NIH Recombinant DNA Research Guidelines and Related Federal Policies," *Recombinant DNA Technical Bulletin* 5 (June 1982): 59–79.) In fact, the search for resolution of these issues led to the identification of many of the future policy questions with which we still wrestle today. Yet it is the lessons we learned from this experience that I especially want to note: the importance of openness, public participation, and the development of a public record in governmental decision-making. They are also the lessons I learned from my clerkship with Judge David Bazelon, whose companion article on this subject appears in this issue. For it was adherence to those standards that helped us deal successfully with Congress (through legislative and oversight hearings); the courts (through suits under the National Environmental Policy Act); and the rigorous oversight of the Department of Health and Human Services and the White House — as well as the press and the public.

Federal Interagency Action

Shortly after the release of the Guidelines in July 1976, President Gerald Ford received a letter from Senators Edward M. Kennedy and Jacob K. Javits urging consideration of measures that would implement the Guidelines, wherever possible, by executive directive and/or rulemaking. They asked the President also to explore every possible mechanism to ensure compliance by the entire research community — including the private sector in this country and public and private sectors internationally — requesting that the President expedite any legislative proposals he might feel required. They concluded by noting that this was a unprecedented issue in the area of biomedical research, one that had been likened in importance to the discovery of nuclear fission.[2]

In September President Ford informed the senators that their concerns could best be addressed by creation of an interagency committee to review the activities of all government agencies conducting or supporting recombinant DNA research, or having regulatory authority relevant to this scientific field. Thus, he directed then-Health, Education, and Welfare Secretary David Mathews to convene, under Presidential mandate, the Federal Interagency Committee on Recombinant DNA Research.[3] The committee's charge included three tasks: to review the nature and scope of recombinant DNA activities in the federal and private sectors; to determine the feasibility of extending the NIH Guidelines to cover those sectors; and to recommend, if necessary, appropriate legislative or executive action. NIH Director Donald Fredrickson served as the committee's chair, and I as executive secretary. Its members represented every federal agency that supported or conducted recombinant DNA research, and all agencies agreed to abide by the NIH Guidelines.

A subcommittee of that interagency committee sought to determine whether existing laws and regulations could be invoked to extend the Guidelines nationally. It concluded that existing laws permitted imposition of some regulatory requirements on a majority of recombinant DNA research, but that there was no single legal authority or combination of authorities that would clearly cover all uses of recombinant DNA techniques.[4] In reaching this conclusion, the subcommittee reviewed several laws that seemed most deserving of detailed consideration:

- The Occupational Safety and Health Act of 1970 (Public Law 91-596), which gives the Occupational Safety and Health Administration (OSHA) broad powers to require employers to provide a safe workplace for their employees. The term "employer," however, is defined in this Act to exclude states and their political subdivisions unless the OSHA standards are voluntarily adopted. Twenty-four states had adopted the standards, but 26 were not subject to them. Furthermore, the standards did not cover self-employed persons. For these reasons, it was determined that OSHA could not regulate all recombinant DNA research.

- The Toxic Substances Control Act (Public Law 94-469), which directs the Environmental Protection Agency to control chemicals that may present an "unreasonable risk of injury to the health or the environment." The subcommittee determined that the materials used in recombinant DNA research appeared to be covered in most cases by the Act's definition of "chemical substance." Section 5 of the Act, however, explicitly exempts from registration chemical substances used in small quantities for scientific experimentation or analysis. This was a most serious deficiency, as registration of activities was thought to be an essential element of any regulatory effort. Furthermore, to meet the specifications of the Act, recombinant DNA research would have to be found to present "an unreasonable risk of injury to health or the environment."

- The Hazardous Materials Transportation Act (Public Law 93-633) and Section 361 of the Public Health Service Act (42 U.S.C. Sec. 264), which give the Department of Transportation (DOT) and the Centers for Disease Control (CDC), respectively, authority to regulate the shipment of hazardous materials in interstate commerce. In implementing these acts with respect to biological products, both the DOT and CDC had essentially aimed at imposing labeling, packaging, and shipping requirements. Thus, these acts were found wanting for regulation of all recombinant DNA research.

On releasing the subcommittee's full report on March 16, then-HEW Secretary Joseph Califano, Jr., commented:[5]

Legislation in this area would represent an unusual regulation of activities affecting basic science, but the potential hazards posed by recombinant DNA techniques warrant such a step at this time. . . . I believe that such a measure is necessary not just to safeguard the public but also to assure the continuation of basic research in this vital scientific area.

Legislative Action

An administration bill based on the committee's recommendations was introduced in the Senate by Edward M. Kennedy, chairman of the Subcommittee on Health and Scientific Research of the Senate Committee on Human Resources, and in the House by Paul G. Rogers, chairman of the Subcommittee on Health and the Environment of the House Committee on Interstate and Foreign Commerce.[6]

Both the House and Senate held hearings to consider the administration Bill and others on the subject. After several redrafts by subcommittees, Senate and House versions of the administration bill were reported out in June. Much of the proposed legislation was more restrictive than the original administration proposal, reflecting the concern of some legislators about the potential risks of recombinant DNA research. However, as Senator Harrison Schmitt notes in his article in this issue, no legislation did pass in that session of Congress, in part because the latest scientific information had demonstrated the safety of much of the research work, and major revision of the Guidelines was under way at NIH and in the Department of Health, Education, and Welfare.

Congressional activity on this matter was extensive at first, but it declined over time. In the 95th Congress, first session (1977), eleven bills for regulation of recombinant DNA research were introduced in the House and five in the Senate. In addition, there were 21 hearings and markups in the House; five in the Senate. In the second session (1978) two bills were introduced in the House, none in the Senate. No hearing was held. In the 96th Congress (1978-80), one bill was introduced (Senate), and one hearing was held on that bill. No bills were introduced in the 97th Congress (1980-82). Therefore, after six years of congressional oversight one concludes that it might not always be necessary — even with a new technology posing both potential benefits and risk — for Congress to pass legislation. In so acting, Congress did not abdicate its responsibility. In fact, it continues to provide oversight — as will be explained in the discussion below on ethical considerations.

Absent legislation, the NIH gave consideration to the need for invoking existing regulatory authorities to cover recombinant DNA research-steps that in the end were not taken. We at NIH found that the guideline model was a more effective mechanism for such a rapidly changing research area, for Guidelines can be modified more quickly than statutes or regulations.

NIH's Recombinant DNA Research Guidelines are especially effective because the scheme respects the roles and responsibilities of the local institutions, although there are established and ongoing procedures to govern revisions at the federal level. The Institutional Biosafety Committee (IBC) model is based on NIH's Institutional Review Boards (IRBs), which oversee health research to ensure compliance with ethical standards in human studies. Those locally derived IRBs administer appropriate oversight within each community; the IBCs provide comparable administration for occupational, environmental, and public health and safety. As Dr. Fredrickson noted: "No law, inspection force, or other external regulation can protect the public interest like responsible and responsive self-governance."[7]

The Guidelines' implementation also demonstrated that collaboration can be achieved between research institutions and regulatory agencies, especially in the transition period as research goes through established regulatory mechanisms and toward product development. NIH was in an awkward position with regard to recombinant DNA research: it both supported the research and provided oversight for it. We had to rely on all parties in the research community — from the industrial, academic, and governmental sectors — to honor these standards. To the credit of all involved, there were few infractions. Those that did occur were corrected within the framework of the Guidelines.

Today the process is moving rapidly from research to application, which is at the outermost boundary of the NIH and enters the province of the regulatory agencies. Yet it is still proceeding smoothly. In part this success results from the ongoing collaboration of all federal agencies on the Recombinant DNA Advisory Committee, a broad-based advisory and technical group, and the inter-agency committee. And it should also be noted that both bodies allow for a combination of scientific and public participation.

Patenting Recombinant DNA Research Inventions

Shortly before the release of the Guidelines, Robert M. Rosenzweig, Vice President for Public Affairs at Stanford University, asked NIH to review HEW policies relating to the patenting of recombinant DNA research inventions.[8] Dr. Rosenzweig pointed out that both Stanford and the University of California were seeking patent protection for recombinant DNA research inventions developed under NIH support by their investigators. In view of the intense public interest in this research, both universities felt the need for a formal advisory opinion by NIH. Indeed, a number of universities were interested in the official view of NIH.

The Department, and therefore the NIH, had two options for dealing with patents developed with departmental funds. The Department could enter into an Institutional Patent Agreement (IPA) with a university or other nonprofit organization that had set mechanisms for administering patents or inventions, thereby granting a general governmental waiver for all patents. Or, for those institutions or organizations not eligible for (or without) a patent agreement with the Department, determination of ownership could generally be deferred until a patent was issued. At that time an institution could petition the Department for ownership of the patent. If the petition were granted, the institution would be required to submit a development plan. (Approximately 90% of all such petitions were granted after presentation of a satisfactory plan for developing or licensing by the institution.)

The NIH completed analysis of public comments on this issue in November 1977 and presented a report to the Federal Interagency Committee on Recombinant DNA Research.[9] All but one agency on the committee agreed that recombinant DNA research inventions should be handled like other inventions. It was the Department of Justice, invoking the nuclear analogy, that argued that the public interest required the government to own all such inventions.

Dr. Fredrickson decided that recombinant DNA research inventions developed under HEW-NIH support should continue to be administered within current HEW patent agreements with the universities. But he felt such agreements should be amended to ensure that—in any production or use of recombinant DNA molecules—the licensees would comply with the physical and biological containment standards set forth in the Guidelines. That decision was announced in March 1978 with the concurrence of the HEW Office of General Counsel and the Public Health Service. Dr. Rosenzweig was informed that Stanford could proceed with its patent application. The patent has been issued, and licenses are being assigned to biotechnology companies.

After many years of public debate over how to enhance use of the results of government-funded research, the last Congress enacted PL 96-517, "The Patent and Trademark Amendments Act of 1980," which gives universities, nonprofit organizations, and small businesses the first right of refusal to title of inventions they make while performing under government grants or contracts (subject to some limited exceptions). In creating this right to ownership, the Act abolishes approximately 26 conflicting statutory and administrative policies. It should be noted, however, that the Act explicitly retains the status quo for contractors other than small businesses, universities, and nonprofit organizations. We can expect a continuing move in the Congress to give all businesses the right of first refusal to invention titles, especially with the *Chakrabarty* decision as an added stimulus (allowing patent protection for recombinant DNA inventions). Full discussion of these issues will appear in an article on patent law in the next issue in this series.

Also of importance in patent policy formulation and implementation is a report from the NIH Director's Advisory Committee, *Cooperative Research Relationships With Industry, The National Institutes of Health Patent Policy Initiatives.*[11] It offers specific guidance concerning allocation of patent rights, reporting requirements for inventions, protection of proprietary information, "march-in" rights, and circumstances when the NIH retains title.

Environmental Impact Statement

One tangential but significant outcome of the Guidelines' development was that the NIH filed an Environmental Impact Statement (EIS) on the Guidelines — a first because EIS's had never before involved basic research. Although it took a year and a half between the draft and the final EIS, that effort provided the courts with an important public record, useful as a basis for decisions when suits arose to enjoin NIH's recombinant DNA activities. After having published the EIS, the NIH, when considering changes to the Guidelines, was able to conduct environmental impact assessments on those changes in the Guidelines without invoking the fullblown EIS process. This is but another example of the flexible implementation the Guidelines themselves allowed.[12]

International Responses to NIH Guidelines

During this period of debate in the United States, many nations were simultaneously reviewing recombinant DNA activities to determine the measures they should consider for safety. With the urging of regional and international bodies, most adopted the US (NIH) or United Kingdom guidelines as a basic policy and procedure framework.

In the United Kingdom, scientific and governmental activities comparable with those in the United States had been under way since 1974, and 1976 guidelines similar to those of the NIH were in place.[13] In Canada a special committee of the Canadian Medical Research Council recommended guidelines to govern the handling of recombinant DNA molecules in Council-supported research. Those guidelines were adopted in February 1977.

The European Molecular Biology Organization, the European Science Foundation, and the European Medical Research Councils were instrumental in coordinating recombinant DNA research activities in western Europe. The three groups worked closely to promote a commonality of safety practices and procedures. With the support of these organizations, technical committees were created in the western European countries to serve as foci for coordinating and monitoring recombinant DNA activities. Many of these bodies functioned in the manner of the United Kingdom's Genetic Manipulation Advisory Group, which is responsible for reviewing all recombinant DNA research to ensure that the projects conform to appropriate safety standards and practices. In addition, international scientific organizations were involved in recombinant DNA activities, including the International Council of Scientific Unions (ICSU) and the World Health Organization. The ICSU Committee on Genetic Experimentation (COGENE), established in October 1976, created task forces to study and analyze various aspects of recombinant DNA research.

Therefore, a November 1977 report of the federal interagency committee concluded that scientific organizations thoughout the world had developed safety standards comparable with those in the United States.[14] But implementation of those procedures was a problem for other countries as it was for the United States: what measures were necessary to ensure that the standards would be observed in both the public and private sectors, and how could one monitor this research? Also critically important was the issue that had been raised at Asilomar and in the US Congress: whether recombinant DNA research might ultimately be used for biological warfare. To the latter, James L. Malone, general counsel of the US Arms Control and Disarmament Agency, stated on July 3, 1975, his opinion that the Biological Weapons Convention prohibits use of recombinant DNA molecules for offensive weapons. And in a statement to the Conference of the Committee of Disarmament on August 17, 1976, Ambassador Joseph Martin, Jr., confirmed that US interpretation of the convention.[15]

Ethical Considerations

As the Guidelines developed through these processes, we also learned the importance of dealing separately with individual, but related, issues. The appropriate emphasis for NIH during the early development of the Guidelines was on the basic research itself and the potential risks involved as industry developed applications.

Yet there remains the longer-term issue of potentially altering the genetic character of man. As human applications become feasible, ethical studies will become essential. The President's Commission for the Study of Ethical Problems in Medicine and Biomedical and Behavioral Research provided a valuable study of the ethical and social implications of genetic engineering, which will contribute greatly to the public policy debates regarding recombinant DNA research.[16] At the time of the guidelines' development there was no need for a presidential commission or an external body to do the essential work, for an executive agency is better equipped to handle such specific problems. This fact was well demonstrated throughout the development and subsequent "deregulation" of the NIH Recombinant DNA Research Guidelines under congressional oversight. The commission went further, however, noting in last November's report that research in the field should continue, but recommending that some mechanism for oversight be established at a national level. [Senator Schmitt's companion article addresses this debate in greater detail.]

The commercialization of science in the biotechnology area as well as the ethical implications of advances in genetic engineering have been the subject of several hearings during the just recessed 97th Congress. Albert Gore, Chairman of the Subcommittee on Investigations and Oversight of the House Committee on Science and Technology, held the latest hearing last November, with many witnesses testifying to identify and discuss the major issues associated with the application of genetic engineering to humans. Included for discussion were the scientific state of the art, as well as the religious, ethical and societal implications of human gene therapy. It was also at that hearing that the Presidential Commission made public its recommendations. Representative Gore there stated his interest in legislation to develop a mechanism for federal review of these broader public and ethical issues. And a recent *New York Times* editorial, commenting on that report, recommended further study of the issues surrounding applications of genetic engineering to fundamental changes in the genetic line of humans.[17] [Later articles in this series will examine these issues in greater detail.]

In sum, the NIH recombinant DNA research guidelines achieved their purpose. Indeed, a *Washington Post* editorial on the guidelines ended with the following statement:[18]

Despite their flaws, the recombinant DNA guidelines have been the model of a responsible approach to a dangerous technology, and of cooperative action between government and the private sector. Had nuclear engineers, pesticide chemists, and numerous others acted with similar caution and sense of public responsibility, everyone would have been much better off.

Cooperative Biotechnology Programs Among Government, Universities, and Industries

Recombinant DNA research activities are now approaching commercialization. The biotechnology industry, therefore, has become more interested in federal research policies and programs that might spur its own goals. Dr. Fredrickson called this technology among others a "revolution" in biology—a flood of basic discoveries in biochemistry, physiology, and medicine, combined with achievements in related disciplines.[19] Today, the biotechnology industry stands ready to work as a partner with governmental and academic researchers to move laboratory innovation toward everyday life.

The NIH's internal planning efforts were a prime ingredient in development of the larger strategies to address today's overriding challenge: how to sustain the capacity and effectiveness of our present research system in a time of limited resources. One such strategy was the stabilization of investigator-initiated research grants, which were proposed to be stabilized at 5,000 new and competing renewal grants out of a total of 16,000 funded each year (approximately 50% of the NIH budget).[20] A complementary strategy rose from our recognition of the in-

terdependence of scientific discovery with the maintence of a pool of well-trained
investigators: the initiative to stabilize the number of research trainees at approximately 10,000 full-time equivalents. Fiscal constraints are today's reality, but the
NIH has done its best to protect the nation's vital investment in scientific research.
Hans Stetten, the Senior Scientific Advisor to the NIH Director, noted at a recent
dinner celebrating the 20th anniversary of the National Institute of General
Medical Sciences that the NIH mission is to conduct and support the best research
in the world. Recombinant DNA technology is an example of the benefit that can
come from this competitive system of investigator-initiated research. That system
must remain a key priority.

These strategies also prompt consideration of related and important issues. We
know that NIH's research should:

- Increase the effectiveness of assessment of clinically relevant research findings, transferring them promptly to health practice;

- Clarify and strengthen the relevance of health research to national needs for
  disease prevention and health promotion;

- Strengthen the government-university partnership in health research
  through policy and procedural changes, especially on selected cost and accountability issues; and

- Bring industry further into the government-university partnership in health
  research—especially in the biotechnology arena, which offers great promise
  for increasing scientific productivity.

These are all major priorities for current NIH Director James Wyngaarden, but
my focus will be on the last item. The strategies at the NIH were based on the
recognition of how most effectively to use governmental research resources to maintain the momentum in the revolution in biology, as exemplified by recombinant
DNA technology. Critical in these considerations was the appropriate role for the
government, universities, and industry.

Government Programs: Federal

Whatever cooperative relationships evolve will reflect existing federal programs and
priorities. Certainly, greater interaction among government, universities, and industries is an essential key to enhanced productivity. Many hearings probing such
collaboration at the NIH and the National Science Foundation note that fact. As I
spoke with program officers in the federal, state, and local governments—as well as
some international ones—I was able to develop a profile on how such interaction
might really work in the biotechnology industry. What follows is a review of those
initiatives, as well as concluding remarks on the university-industry relationships
themselves.

The National Institutes of Health has been a primary supporter of recombinant
DNA research and related technologies over the past several decades. Cumula-

tively, investments are in the billions with direct results in the development and application of this technology. Dr. Richard Krause looked at the excitement and reality those billions will bring in an opening article in this series.[21]

Funding for "Genetic Manipulation"

Research involving the general area of "genetic manipulation" has received steadily increasing NIH funding since 1978: from $61 million for 554 projects in 1978 to $184 million for 1,523 projects in fiscal year 1982.[22] NIH further anticipates that the use of recombinant DNA techniques in general will continue to increase in research projects. Support for research in monoclonal antibodies, hybridomas and cell-fusion technology in 1980 amounted to 553 awards for $48 million. These dollars have done more than bring in new drugs, vaccines, and food additives. Recombinant DNA technology as a standard research tool now allows a new understanding of the molecular bases of development, of evolution, and of oncogenesis. And we are approaching the time when molecular biology will be linked with sophisticated data-processing technologies. In the medical arena, there is an improved understanding of genetic diseases, as we can now relate action at the molecular level to events at the organic level—or even at the level of the organism itself—using that information for clinical diagnois to tailor drug therapy to the individual.

The National Science Foundation has also been a leader in support for genetic engineering research.[23] In addition to grant and contract monies there are special programs with a biotechnology emphasis under the Directorate for Scientific, Technological, and International Affairs—as well as their industry-university cooperative research centers and projects. There has been an NSF planning grant to the University of Wisconsin for a center for hybridoma research; and among the university–industry projects for fiscal year 1982 are development of continuous mixed culture processes, search for new restriction endonucleases, and the study of the structure and function of *Bacillus thuringiensis*.

In 1977 the National Science Foundation began a Small Business Innovation Research (SBIR) grant program to encourage small business participation in NSF programs. In the summer of 1982, Congress passed the Small Business Innovation Development Act (P.L. 97-219), modeled on the NSF program. The Act seeks to stimulate technological innovation, use small business to meet federal research and development needs, and increase private sector commercialization of innovations derived from federal research and development. The Act mandates federal agencies to establish SBIR programs if their fiscal year 1982 extramural budgets for research and development exceed $100 million at the following levels: .20% for fiscal year 1983, .60% for 1984, 1% for 1985, and 1.25% for fiscal years 1986–88. Agencies with lower R&D budgets must establish specific goals to encourage participation by small business through contracts, grants, or cooperative R&D agreements.

As in the NSF program, the SBIR research will be funded in three phases. Phase I involves experimental or theoretical research, or R&D efforts on prescribed agency requirements, with funding up to $50,000 over six months. Phase II is the principal research or R&D effort; funding is based on the results of phase I and the scientific and technical merit of phase II. Awards will range from $200,000 to $400,000 over two years. Phase III is to be conducted by the small business (including joint ventures or R&D partnerships), which will pursue with nonfederal funds commercial applications of the research or R&D funded in phases I and II.

The Small Business Administration, in a November 1982 policy directive, provides the guidance for the programs that will be developed under legislative mandate.[24] All agencies are now creating the program instruments and solicitations for small businesses.

Vigorous Support Programs

In the past several years, several other departments and agencies have begun vigorous support programs for recombinant DNA research and related biotechnologies. The Department of Agriculture, the Department of Energy, and the Environmental Protection Agency, National Aeronautics and Space Agency, and the Department of Defense are all looking at applications relevant to their areas of interest. For example, Agriculture's grant support for fiscal year 1981 and 1982 amounted to $1.8 million and $3.5 million, respectively.[25]

The Department of Energy is particularly interested in genetic engineering techniques for biomass energy R&D. DOE's biological energy research program was established to conduct fundamental studies in biology oriented toward energy conversion and conservation and to underpin future developments of energy-related biotechnology. Their focus is on plant and microbial sciences in such areas as plant genetics, fermentations, and other bioconversions. Especial emphasis is given research in genetics, biochemistry, and physiology aimed at defining feasible targets for genetic engineering in plants and microorganisms. The objectives are to identify mechanisms of expression and important traits which might be susceptible to genetic manipulations. The Biomass Energy Technology Division focuses on biomass energy R&D.[26] And, indeed, the processes for refining biomass from trees or from agriculture residues are fast approaching commercialization.[27]

The Environmental Protection Agency over the past three years has provided support for risk assessment programs in recombinant DNA and related technologies, including future environmental trends and problems in the agricultural and industrial use of applied genetics and biotechnologies.[28] The EPA has also conducted three workshops on an Environmental Risk Assessment Program:[29] (1) on public health concerns—advice to the agency on a focus for sewage and aerosol programs; (2) on environmental impact—emphasizing problems associated with survival and dispersion of novel genomes; and (3) on exploitation of recombinant DNA technology—stressing the importance of the need for basic research prior to attempted application. Research grants are being given in the area of aerosols and sewage, and aerosols in environmental genetic exchange. Finally, there have been special studies on physical containment controls, with concomitant research on risk assessment programs regarding deliberate environmental release, gene transfer, and genetic stability.[30]

The *National Aeronautics and Space Administration's* Life Science Division supports biologic sciences research in space, much of it relating to the effects of zero gravity on animals and humans. Some are looking into "exobiology" — studies of the origins of life. Exobiology, in fact, involves fundamental research in DNA and RNA sequencing for purposes of studying evolution, but this NASA program does not yet use recombinant DNA technology. Another program within the life sciences division is on controlled ecological life support systems (CELSS). NASA has issued a number of reports on this program, including one on genetic engineering applications — especially for breeding plants to optimum growth in space.[31]

The Department of Defense also has several biotechnology research initiatives under way.[32] The Army's primary interest is in medical measures: vaccines, toxins, drugs, diagnostic systems, and hazardous organisms, with a particular interest of the US Army Medical Research Institute of Infectious Diseases in the development of antiviral drugs and vaccines. The US Army Medical Research Institute of Chemical Defense has program interest in biotechnology applications for detection and defense against chemical weapons, and Army's Chemical Systems laboratories are supporting the development of monoclonal antibodies for use in detection of chemical agents and toxins. DOD's Defense Applied Research Projects Agency (DARPA) has a series of biotechnology initiatives under way in the area of chemical ultrasensors and production of biopolymers. The Office of Naval Research, its Naval Research Laboratories, and the Naval Air Systems Command are creating a biotechnology research program for developing new materials, new films (electronics, optics, etc.), and new detecting devices. The Office of Naval Research and Naval Air Command are sponsoring a series of biotechnology workshops at the National Academy of Sciences during 1983–84 in biomolecular electronics, material sciences, and microecology (biofilm coatings, drag reduction polymers, and detection devices). These NAS workshops follow those sponsored by ONR at the North Carolina Biotechnology Center in the fall of 1982.

Relevance to Needs

Finally, there are the programs of specific relevance to felt needs, such as those under way at the Agency for International Development, the Veterans' Administration, and the National Bureau of Standards. The *Agency for International Development* is in the process of initiating new efforts for biotechnology program support. AID sponsored a workshop at the National Academy of Sciences on "Priorities in Biotechnology Research for International Development" in 1982. Panels looked at monoclonal antibodies, animal production and embryo transfer technology, vaccine development, plant cell and tissue culture implants, biological nitrogen fixation, and conversion of biomass to fuel, feeds, and other useful chemicals.[33] Specific priorities of developing countries may differ, but a major goal is to use biotechnology as a key means for national growth and development, primarily through application of genetic engineering techniques to agriculture, health, and energy.[34]

The National Bureau of Standards recently issued a planning report regarding the bureau and measurement-related needs associated with industrial biotechnology.

That report examines the organic chemical industry, identifies technologies important to industrial bioprocesses, and evaluates bureau research and service capabilities related to measurement problems in industrial biotechnology, particularly in the Centers for Chemical Physics, Analytical Chemistry, and Materials Science. Outlined in the report are several areas where NBS could initiate activities to provide measurement-related research and service infrastructure support for industry needs in biotechnology.[35]

How much industry-university-government collaboration actually develops from these initiatives depends in large part on the success of the programs themselves. Last year, a National Bureau of Standards planning report examined the federal government's role in encouraging industry–university cooperative research, reviewing the NSF grant program for industry–university cooperative research projects, initiated in 1978.[36] This program has attempted to sustain scientific and engineering knowledge essential for future technological innovation, having awarded by the end of its first four years 231 competitive grants totaling nearly $30 million. Since projects supported by this program tend to be typically larger than other NSF competitive research grants, some of the leading R&D-performing companies in the United States are seeking funding from this program.

The report concludes that the program has been successful in linking companies and universities and in increasing the potential for technology transfer, although it is difficult to judge whether any *new* linkages have been forged. The report does note that the program seems to contribute to increasing the relevance of NSF's academic research, a significant fact since this program commits more than 1% of NSF's current research budget. The report notes that such money would go through the normal competitive grant process without this program, but giving up the program might mean lessening NSF's commitment to the nation's long-term economic development. Finally, the report outlines some alternative federal policies to promote collaboration, including patent policies, the Economic Recovery Tax Act (providing incentives for companies to support university research), and various antitrust policies that would promote collaboration and cooperation.

The SBIR program, too, is being watched closely for its effect on future trends. Dr. Donald Kennedy, in the first issue of this *Technology In Society* series, takes exception to this act and similar programs, noting what he feels is a misunderstanding of the fundamental relationships for R&D among government, university, and industry.[37] Senator Warren Rudman, sponsor of the legislation, argues to the contrary—that this program will invigorate the US industrial base and enhance R&D productivity. Implementation of the program and evaluation over the next three to five years should provide a good test of whether such a program can meet Senator Rudman's goals, or whether Dr. Kennedy is correct.

As Dr. J. Leslie Glick outlines in his article in the previous issue of this journal, the industrial impact of recombinant DNA technology will be broad, touching virtually all the research areas sponsored or developed by the above federal agencies.[38] Today there are perhaps 200 young firms specializing in biotechnology. Their total capitalization exceeds $700 million. In addition, there are many larger companies now actively involved in biotechnology, spending probably another $5 billion of internal R&D funds.

Although the interests of all the partners in this research are not similar, they are common. Small genetic engineering companies can focus various talents to invigorate the productivity of our technological base, and there is promising collaboration between smaller and larger companies in the areas of health, energy, and agriculture. Further, the Small Business Innovation Research Grant program in such agencies as NIH, NSF, DOE, and Agriculture should prompt active and fruitful interaction. Over the next decade, through stimulation and collaboration in terms of both policy and dollars, we should see greater R&D results among all the partners—with industrial application involving both small and large firms.

Governmental Programs Abroad

There is considerable international interest, too, in the applications of recombinant DNA technology and the related biotechnologies. In discussions with science attaches of the foreign embassies, many report biotechnology initiatives in their respective countries.[39] France has launched several technology initiatives with the highest priority for biotechnology. Further, the United Nations Industrial Development Organization (UNIDO) is considering the establishment of an international center for genetic engineering and biotechnology, as recommended at a 1981 meeting convened by the UNIDO Secretariat. After a review of research and development priorities, especially regarding the potential of biotechnology to the needs of lesser developed countries, UNIDO's 16 experts noted the importance of such a move.[40] Areas of research interest are development of human and animal vaccines against parasites, bacteria, and viruses that cause the more important infectious disease problems in lesser developed countries.[41]

Another UNIDO emphasis is energy and fertilizer production, with particular interest in conversion of cellulose to ethanol, production of fertilizer, enhancement of nitrogen fixation, and other methods of improving crop productivity.[42] A complementary strategy is to improve agricultural and food products by novel methods of plant breeding and tissue culture, using such means as clonal reproduction, biological pesticides, photosynthesis, or other complex plant processes.[43] Other proposed programs refer to tertiary oil recovery from petroleum wells by microbial means, enhancing recovery from wells already pumped of the bulk of their oil.[44] These strategies point out the relevance of comments by Dr. Nyle Brady, AID's senior assistant administrator for science and technology. He noted in *Science* that genetic engineering will probably be a primary mechanism to achieve the goal of increasing food production in areas less well endowed with natural and economic resources, although chemistry and chemicals will also play a vital role.[45]

The World Bank is also looking at biotechnology. The Consultative Group on International Agricultural Research, which the bank chairs, set forth in a 1981 report potential applications of tissue culture and genetic engineering relating to the work of the international agricultural research centers it supports. That report focuses on biological nitrogen fertilization.[46] In another report prepared for Dr.

Charles Weiss, the World Bank Science and Technology Advisor, Dr. Christian Orrego reviewed the potential applications of biotechnology to the needs of lesser
developed countries in fuel production, agriculture, and forestry. Dr. Orrego cites
many of the areas included in the UNIDO reports. He recommends that the World
Bank evaluate the potential effects of the new biotechnologies before their introduction, as well as reviewing the same questions after introduction.[47] But
Weiss's perspectives are also relevant: he asks that innovation be promoted in areas
of need, whether the commercial demand is great or not, citing as a successful example the international agricultural research institutes.[48]

No review of governmental initiatives and the anticipated international applications can ignore discussions regarding potential US export controls on high technology products. Currently, the President's Office of Science and Technology Policy,
headed by George A. Keyworth II, has conducted a study on biotechnology and
technology transfer. An interagency work group on biotechnology reported on the
commercial and national security aspects of the transfer abroad of genetics and biotechnology advances and assessed the need for government policy to sustain US scientific and technological leadership in biotechnology and to describe and evaluate
policy options. That report recommends improvement, streamlining, and updating the current system of monitoring and control, but states that the situation does
not now warrant more restrictive controls.

International Scientific Exchange

We in the biotechnology industry are heavily dependent on international scientific
exchange, in terms both of industrial applications and the free exchange of
technical information through scientific meetings and publications. The Center for
Strategic and International Studies of Georgetown University held a conference last
December on US strategies and foreign industrial targeting, where there was
general agreement among the participants that high technology would be an important component of international trade in the 1980s.[49] This conference, first in a
series under the center's newly established Technology and International Business
Program, examined the future course of US trade policy and assessed the options
for domestic initiatives in such areas as tax, antitrust, and manpower policies. That
meeting will be followed by the formation of small working groups to define US
strategies on key issues selected from the conference agenda. The Georgetown
Center is but one of several governmental and private sector initiatives in this area,
which itself looms as a key priority for government in the 1980s.

Of special importance to government policy-making is a forthcoming (fall, 1983)
study by Congress's Office of Technology Assessment that will provide a comparative assessment of the commercial development of biotechnology in the United
States and abroad.[50] This assessment is examining whether biotechnology is
developing in the United States in such a way that this nation is likely to be in a
competitive position with other nations in the years ahead. Besides describing the
state-of-the-art in this and other countries, major influences likely to determine
future development of the industry are being reviewed. These include government
policy on funding of research, patents, health and safety regulations, antitrust
laws, and taxation as well as industrial-academic relationships and their influences
on funding, research, manpower, training, and information flow.

Governmental Programs in the States

There is also enormous ferment in the states, specifically in regard to the efforts of many of them to support biotechnology programs as a means of bringing high technology industry within their borders. The National Governors' Association has a task force of 20 governors devoted to coordinating and promoting such technology through state support, tax benefits, and stimulation of university-industry connections. Their report details many state programs, including public-private forums for discussing technological innovation, promoting resource efficiency, linking university and research development with high technology industry, increasing the technical training of personnel, and financing technological innovation in new and different ways.[51] In addition, the California Commission on Industrial Innovation recently issued a report and a series of recommendations that seek to promote a new industrial strategy for innovation at the national and state levels.[52]

Complementing such state efforts is a report by the Joint Economic Committee of the Congress, which surveyed several hundred high technology firms in all 50 states regarding the factors that led them to make a location decision.[53] That report notes that California's Silicon Valley and Boston's Route 128 were industrially created — no state planning encouraged their growth. However, North California conceived its Research Triangle Park from state initiatives 20 years ago. The Joint Committee report documents the considerable time and effort that led to what is now perceived as a successful payoff. I, too, have spoken with a number of persons involved in this North Carolina effort, including Quentin Lindsey of North Carolina's Board of Science and Technology. All of them acknowledged the need for industrial involvement at the earliest planning stages.[54] In fact, North Carolina's biotechnology center, with the goal of strengthening biotechnology in the state, involves both university and industrial representatives in all its program development activities through workshops, university-industry planning committees, and joint funding.[55]

These effective state initiatives hold models for future growth of biotechnology itself and collaborative arrangements in general. Indeed, North Carolina's Governor James B. Hunt, Jr., noted that the center of gravity for technological innovation must shift from the federal government to the state governments, a fact that the Joint Economic Committee report also brought out. It is, therefore, not surprising that several states, including Maryland, now have or are developing similar programs.[56]

University-Industry Collaboration

In my discussion of development of NIH's recombinant DNA research guidelines, I noted that the NIH Director's Public Advisory Committee (DAC) was a major force behind the development of the NIH's strategies. When, in late 1979, that committee looked toward stablizing support for the principal investigator, it also recommended that NIH study various means of enhancing the stability of the investigator's environment: namely, the academic research community. The DAC, therefore, studied in detail potential cooperative relationships among government, universities, and industry as an adjunct to the NIH planning effort, assessing various means to continue the research capacity of each of the partners while ensuring appropriate accountability for the use of public funds. Again in 1980-82, the committee returned its focus to the respective roles of government, universities, and industry in enhancing the university research environment.[57]

Listening to members' discussions at the Director's Advisory Committee, I heard evidence repeatedly presented to defend the view that the appropriate role of the federal government is to support fundamental research at the universities. Industry's role was defined as supportive of applied research and technology transfer. Indeed, these were the conclusions of many meetings on the topic, along with a recognition that industrial support will never be a substitute for heavy federal investment in basic research. Strong confirmation of that position was voiced again at a recent meeting on industry-university cooperation sponsored by the National Science Foundation.[58] One speaker there cited a Battelle study, reporting that for every 1% reduction in federal R&D funds there needs to be a corresponding 20% increase from industry. Certainly, such an increase is unlikely now, when industrial profits and earnings are still shaken by the recession.

It is logical that a number of the issues raised at the Director's Advisory Committee during my tenure at the NIH continue to dominate discussions among government, universities, and the biotechnology industry. Included are questions such as: What should be the appropriate split between federal and private sector for research in this area? What should be the nature of support for clinical research? Can the private sector do such a large share? How can the conflict of interest between university laboratory and industrial laboratory be handled, especially where there are exchanges of investigators? Finally, what policies will govern the allocation of patent rights, licensing, and royalty fees when there is commingling of government and industry funds for projects at universities and nonprofit institutions?

The Continuing Debate

A number of these issues continue to be debated, as they were at a May 1982 meeting of the Industrial Biotechnology Association and by Donald Kennedy and his colleagues from Harvard, the Massachusetts Institute of Technology, the University of California, and the California Institute of Technology at a West Coast gathering at Pajaro Dunes in March 1982. Dr. Kennedy's article in this series summarizes his policy perspectives in light of this meeting.[59] Another conference devoted to university-industry relationships was held last December at the University of Pennsylvania, with attendees from Yale, the University of Michigan, Washington University, Cornell, Johns Hopkins, Princeton, and the University of Texas. Several issues arose from those meetings that point to areas where perspectives might differ among the research partners:

- Program relevance — what industry is proposing versus what the university can do.

- Appropriate time frames — industry's need for faster payoff, and the university's need for longer basic research commitments.

- Protection of proprietary information, patent rights, and appropriate administration of licensing and royalty arrangements.

- Conflict of interest — potentially difficult for the university and its faculty who participate in industrially supported projects or in research corporations created by a university.

The *Washington Post* commented in an editorial on the Pajaro Dunes meeting that "Universities, and science as a whole, would benefit from an attempt to hammer out rules to guide development of new relationships with business that won't endanger academic science."[60] But in a *Science* editorial, Dr. Philip Abelson took note of the differing values in academic and industrial research efforts, while citing the desirability of cooperation.[61] Dr. Abelson voiced caution about the new arrangements between universities and industry that create, in effect, industrial R&D laboratories in university settings. Of special concern to him is the harm that could come to the university's essential functions of undergraduate and graduate education.

Addressing these concerns, A. Bartlett Giamatti, president of Yale University, publicly outlined his institution's plans to answer potential problem situations.[62] Yale University intends to issue a statement of policy governing the nature and extent of university and faculty involvement in commercial application. The policy will rely on openness and free dissemination of ideas, while recognizing the need of profit-oriented companies to treat knowledge as private property. Although the university will continue to allow relationships between faculty members and commercial companies, a faculty member who goes beyond any reasonable definition of "consulting" may be asked to take an unpaid leave of absence, or to sever his or her ties with the university.

The Association of American Universities, in a report on ethical issues in industry-university ties, proposed that it act as a "clearinghouse" to gather and distribute information bearing on policies in the area of industry-university relations. However, that report stopped short of asking a group to promulgate a code of ethics to guide the behavior of the institutions and their faculties because[63]

the conditions exist for intelligent and thoughtful decision-making on these issues at the level of individual institutions . . . [P]olicies that are informed by wide experience and that are tailored to individual circumstances are preferable to injunctions that are broadly enough cast to cover the multitude of local circumstances that exist among our many universities.

Dr. William O. Baker has commented that we should move the present industry-university focus from "who-pays-for-what-contract" into a recognition of what can be done for the future of our society and nation.[64] And in that spirit, some models are developing. Dr. Donald Brown, Director of the Department of Embryology at the Carnegie Institute, established the Life Sciences Research Foundation to seek industrial support for research and training in the life sciences at academic and research institutions through peer-reviewed post-doctoral fellowships. He and his board face tough hurdles, but they have generated enough support to offer several postdoctoral fellowships beginning October 1, 1983.

In preparing his program, Dr. Brown looked carefully at the whole area of collaborative support, studying especially the German chemical industry. There, industry supports research and training, as well as chemistry instruction and improvement in the vocational and professional situation of the chemist. Financing comes from a special membership fee, amounting to .12% of the member firm's gross proceeds from the chemicals business. Total income to the fund from 1950 through 1980 was approximately 160 million Deutschmarks, and its support included training of younger scholars at the universities, as well as a publications program that awards a prize for the best publication in chemistry. Also emphasized are continuing education in chemistry and enhancement of job opportunities.

In the United States, one model now emerging is that of the Semiconductor Research Cooperative, established by our semiconductor industry to increase the level of focused research.[65] The fund resulted from what was perceived as severe and continuous pressure from foreign compeition in an industry believed to be as basic as steel and automobiles were 50 years ago. The cooperative was created in the belief that the world leader in research will win the lion's share of market performance in the future, and that for a number of reasons American semiconductor research efforts are stagnating — or declining. To reverse the trend, the founders believe that an initiative will have to come from outside the government. The SRC is aimed at stimulating and enhancing semiconductor research at universities, as well as thereby increasing the number of graduate students and advance degree graduates. The first year (1982) of operation will distribute approximately $6 million in contracts. 1983 should see an increase to about $11 million. Centers of excellence will ultimately be selected and used to focus activities in broad research areas.

Industry's Increasing Investment

Dr. Frank Press, President of the National Academy of Sciences, recently discussed industry's heavy and increasing investment in research and development. The NSF projects a 12% rise in corporate R&D spending in 1982, and an 11% increase in 1983.[66] Dr. Press recommends a complementary effort by government to support basic research. Indeed, he goes beyond that to suggest that a small percentage of federal development funds be reprogrammed for basic research.[67] I might add that, as President Carter's Science Adviser, Dr. Press strongly supported NIH's commitment for 5,000 new and competing grant awards.

Stephen Jobs, head of Apple Computer, spoke at the National Governors' Association Task Force meeting in February 1982, noting that, until the 1930s, the prevalent US occupation was "farmer." In the 1930s, "laborer" became predominant, remaining so until 1979, when it suddenly switched to "clerk." Jobs noted these milestones in a trail he sees as leading toward a knowledge-and-information-based society, rather than a production-based society. Further, Jobs related these shifts to our current shortage of engineers, caused by what he sees as cannibalization of our universitites. He points out that we're eating our seed corn if you will—all of which makes support such as that by the Semiconductor Research Cooperative more imperative.[68]

At the same meeting Senator John Glenn quoted very much earlier comments by Harry Truman:[69]

Progress in scientific research and development is an indispensable condition to the future welfare and security of the nation. No nation can maintain a position of leadership in the world of today unless it develops to the full its scientific and technological resources. No government adequately meets its responsibility unless it generously and intelligently supports the work of science in university, industry, and in its own laboratories.

The world has changed little in this respect. As we in biotechnology enter the industrial arena, we can and should develop a number of interesting models to enhance government-university-industry interaction. I would hope as the biotechnology industry matures, we, too, can provide support to our academic colleagues and institutions. Our goal should and must be to protect each party's interest, while promoting the US technological lead with effective productivity gains. But success depends on our national commitment to the support of basic research —the pillar of all industrial R&D in this nation. Such support for basic research does not come cheaply, but that is not a new problem: from Ecclesiastes we learn that "A feast is made for laughter, and wine maketh merry; but money answereth all things."

Notes

1. *Federal Register* 41 (1976), pp. 27902–27943.
2. Recombinant DNA Research Documents Relating to "NIH Guidelines for Research Involving Recombinant DNA Molecules," no. 2, pp. 157–160. (Where no other title is given, the number preceding the page numbers refers to that volume in the Recombinant DNA series, as follows; Volume 1 (February 1975–June 1976); Volume 2 (June 1976–November 1977); Volume 3 (November 1977–September 1978); Volume 4 (August–December 1978); Volume 5 (January 1979–January 1980); Volume 6 (January 1980–December 1980); Volume 7 (November 1980–August 1982).
3. 2:161–164.
4. 2:279–345.
5. 2:276–277.
6. 2:559–577.
7. Donald S. Fredrickson, "The Public Governance of Science," Columbia University Bicentennial Lecture, December 9, 1976; published in *Man and Medicine* 3:2 (Winter 1978), pp. 77–78.
8. 2:56–61.
9. 2:2–47.
10. 2:1A–1B.
11. Advisory Committee to the Director, National Institutes of Health, *Cooperative Research Relationships with Industry: National Institutes of Health Patent Policy Initiatives* (Bethesda, MD: National Institutes of Health, 1981); this includes all relevant proceedings of the committee.
12. For a review of the NIH compliance with the National Environment Policy Act of 1969 (NEPA) (42 U.S.C. S 4321) and resultant litigation, see *NIH Recombinant DNA Technical Bulletin* 5 (1982), pp. 65–66.
13. 2:441–460.
14. 2:381–500.
15. 2:486–494.
16. President's Commission for the Study of Ethical Problems in Medicine and Biomedical and Behavioral

Research, "Splicing Life — A Report on the Social and Ethical Issues of Genetic Engineering with Human Beings," November 1982.

17. "The Rules for Reshaping Life," editorial, *The New York Times,* December 29, 1982, p. A18.
18. "DNA: Risks and Guidelines," editorial, *The Washington Post,* February 4, 1980.
19. See D.S. Fredrickson, "Biomedical Research in the 1980's," *New England Journal of Medicine* 304:9 (February 26, 1981), pp. 509–517.
20. See *NIH Research Plan: FY 1983–85* (Bethesda, MD: National Institutes of Health, 1981).
21. R. Krause, "Is the Biological Revolution a Match for the Trinity of Despair?" *Technology In Society* 4:4 (1982), pp. 267–282.
22. James B. Wyngaarden, "Government, Industry and Academia: A Bermuda Triangle?" presented at the National Conference on University-Corporate Relations in Science and Technology, University of Pennsylvania, Philadelphia, Pennsylvania, December 16, 1982; and Michael I. Goldberg, "Biomedical Research: Common Interests," presented at the Industrial Biotechnology Association annual meeting, Washington, DC, October 26, 1982.
23. See O.R. Zaborsky, "Biotechnology at the National Science Foundation" (Washington, DC: National Science Foundation, 1981).
24. *Federal Register* 47 (1982), pp. 52966–52976.
25. See, *e.g.,* Department of Agriculture, "Food and Agriculture Competitively Awarded Research Grants, FY 1981" (November 1981) and "Emerging Biotechnologies in Agriculture: Issues and Policies," Progress Report, (November 1983).
26. See *Annual Report and Summaries of FY 1982 Activities Supported by the Division of Biological Energy Research* (Washington, DC: US Department of Energy, 1982). See also R. Rabson and P. Rogers, "The Role of Fundamental Biological Research in Developing Future Biomass Technologies," *Biomass* 1 (1981), pp. 17–37.
27. H.R. Bungay, "Biomass Refining," *Science* 218 (1982), pp. 643–646.
28. R.H. Zaugg and J.R. Swarz, *Assessment of Future Environmental Trends and Problems: Industrial Use of Applied Genetics and Biotechnologies* (Washington, DC: US Environmental Protection Agency, 1981); and G.H. Kidd, M.R. Davis and P. Esmailzadeh, *Assessment of Future Environmental Trends and Problems: Agricultural Use of Applied Genetics and Biotechnologies* (Washington, DC: US Environmental Protection Agency, 1981).
29. M. Levin, "Review of Environmental Risk Assessment Studies Sponsored by EPA," *NIH Recombinant DNA Technical Bulletin* #5 (1982), pp. 177–180.
30. See L.M. Mallory, J.L. Sinclair, L.N. Liang and M. Alexander, "A Simple and Sensitive Method for Assessing Survival in Environmental Samples of Species Used in Recombinant DNA Research," *NIH Recombinant DNA Technical Bulletin* 5 (1982), pp. 5–6.
31. See E.J. Johnson, "Genetic Engineering Possibilities for CELSS: A Bibliography and Summary of Techniques" (National Aeronautics and Space Administration, March 1982). See also "Controlled Ecological Life Support Systems (CELSS) Program Plan" (National Aeronautics and Space Administration, March 1982); and "Life Science Considerations for Space Station" (National Aeronautics and Space Administration, September 1982).
32. See, *e.g., Fiscal Year 1983 Research and Development Program, A Summary Description, March 30, 1982* (Defense Advanced Research Projects Agency, Department of Defense), pp. 40–41; and The United States Army Medical Research Institute of Infectious Diseases (USAMRIID) Annual Report for Fiscal Year 1982 (Fort Detrick, MD: In press).
33. *Priorities in Biotechnology Research for International Development — Proceedings of a Workshop* (Washington, DC: National Academy of Sciences, 1982).
34. See M.S. Swaminathan, "Biotechnology Research and Third World Agriculture," *Science* 218 (1982), pp. 967–972.
35. T.C. O'Brien, "Planning Report 12; NBS and Industrial Biotechnology: Technical Developments and Future Measurement Needs" (Department of Commerce, National Bureau of Standards, April 1982).
36. R.M. Kunkel, "Planning Report: Encouraging Industry-University Cooperative Research: An Assessment of the Federal Government Role" (Department of Commerce, National Bureau of Standards, April 1982).
37. Donald Kennedy, "The Social Sponsorship of Innovation," *Technology in Society* 4:4 (1982), pp. 253–265.
38. J.L. Glick, "The Industrial Impact of the Biological Revolution," *Technology In Society* 4:4 (1982), pp. 283–293.
39. *Biotechnology: Report of a Joint Working Party, Advisory Council for Applied Research and Development, Advisory Board for the Research Councils, The Royal Society* (London, UK: March 1980). See also *Biotechnology: A Development Plan for Canada — Report of the Task Force on Biotechnology to the Minister of State for Science and Technology* (Ottawa, Canada: February 18, 1981).
40. See *Five-Year Work Program of the International Centre for Genetic Engineering and Biotechnology* (United

Nations Industrial Development Organization, September 1982), and C. Heden, *Selective Application of Advanced Biotechnology for Developing Countries* (UNIDO, September 1982).

41. A. Bukhari and U. Pettersson, *Application of Genetic Engineering and Biotechnology for the Production of Improved Human and Animal Vaccines with Particular Reference to Tropical Diseases* (UNIDO, September 1982).

42. R. Wu, *Application of Genetic Engineering for Energy and Fertilizer Production from Biomass* (UNIDO, September 1982).

43. D. McConnell, *Improved Agricultural and Food Products Through Genetic Engineering and Biotechnology* (UNIDO, September 1982).

44. A. Chakrabarty, *Hydrocarbon Microbiology with Special Reference to Tertiary Oil Recovery from Petroleum Wells* (UNIDO, September 1982).

45. See N.C. Brady, "Chemistry and World Food Supplies," *Science* 218 (1982), pp. 847–853.

46. Consultative Group on International Agricultural Research, *1981 Report on the Consultative Group and the International Agricultural Research It Supports—An Integrated Report* (Washington, DC: CGIAR Secretariat, World Bank, September 15, 1982), pp. 10–14.

47. Christian Orrego, World Bank Science and Technology Report Series, *Evaluation of Microbial Technologies Involved in Fuel Production, Agriculture, and Forestry* (Washington, DC: World Bank, August 1981). For an excellent scientific review of the World Bank's programs, see *Science and Technology in World Bank Operations* (Washington, DC: World Bank, September 1980).

48. C. Weiss, Jr., "Mobilizing Technology for Developing Countries," *Science* 203 (1979), pp. 1083–1089.

49. Georgetown University Center for Strategic and International Studies, a report of a Conference on US Strategies and Foreign Industrial Targeting, December 8, 1982, Washington, DC.

50. Office of Technology Assessment, *Comparative Assessment of the Commercial Development of Biotechnology* (In press).

51. *State Activities to Encourage Technological Innovation: An Update* (Sacramento, CA: Published for the National Governors' Association by the California Commission on Industrial Innovation, 1982).

52. *Winning Technologies: A New Industrial Strategy for California and the Nation,* Report of the California Commission on Industrial Innovation, September 1982.

53. *Location of High Technology Firms and Regional Economic Development—A Staff Study,* Subcommittee on Monetary and Fiscal Policy, Joint Economic Committee, US Congress (Washington, DC: US Government Printing Office, 1982).

54. See the FY '81 Report and the FY '82 Plan of the North Carolina Board of Science and Technology (Raleigh, NC: Office of the Governor, July 1981).

55. Cooperative efforts of this sort are underway in the biomedical, agricultural, and engineering applications of biotechnology (personal communication, Don I. Phillips, Acting Director, North Carolina Biotechnology Center).

56. *A Report of the Governor's Ad Hoc Committee on High Technology,* submitted to Harry Hughes, Governor of Maryland (October, 1982).

57. NIH patent policy initiatives, *op. cit.* at note 11.

58. 1982 Conference on Industrial Science and Technological Innovation, "New Paths to Scientific and Technological Innovation," May 1982 (National Science Foundation, Washington, D.C., 1982).

59. Kennedy, op. cit.

60. Editorial, "Uneasy Partners," *Washington Post,* April 6, 1982.

61. P. Abelson, editorial, "Differing Values in Academia and Industry," *Science* 217 (1982): 1095.

62. A. Bartlett Giamatti, "The University, Industry, and Cooperative Research," *Science* 218 (December 24, 1982): 1278.

63. Report of Association of American Universities on the ethical considerations arising from university-industry interests, October 28, 1982. Contained in letters to Rep. Don Fuqua and Rep. Albert Gore, U.S. Congress.

64. Personal communication, January 17, 1983, W.O. Baker, Bell Laboratories, Murray Hill, New Jersey 07974.

65. For industrial support of microelectronics centers at universities such as Stanford, see C. Norman, "Electronics Firms Plug into the Universities," News and Comment, *Science* 217 (1982): 511–514. See also, P. Abelson, Editorial, "Leadership in Computer Technology," *Science* 219 (1983).

66. Science Resources Studies, NSF 82–311, June 1982.

67. F. Press, "Rethinking Science Policy," *Science* 218 (1982): 28–30.

68. National Governors Association Task Force on Technological Innovation," "Utilizing America's Technological Resources: New Challenges to the States," (National Governors' Association, Washington, D.C., February 1982), pp. 14–15.

69. *Ibid.,* pp. 15–19.

Part IV

11

Introduction

Joseph G. Perpich

Having opened with perspectives on research and development programs now present in universities, government agencies, and industry itself, the book next examined the impact of biotechnology on the workings of government: Congress, the courts, and the executive branch. Part IV widens the discussion, examining the impact of biotechnology on corporate, regulatory, and patent law. Also noted are the evolving legal relationships among universities and industry as well as the impact of the government on that relationship.

Harold P. Green

The first contributor to this section is Harold P. Green, Professor of Law and Associate Dean for Post-J.D. Studies at The George Washington University National Law Center. Mr. Green holds an undergraduate degree in economics from the University of Chicago and is a graduate of their law school. He served in the Office of the General Counsel, US Atomic Energy Commission from 1950 to 1954. He was a partner in the Washington, D.C., office of Fried, Frank, Harris, Shriver, and Kampelman until 1983. From 1964 to 1979 he was also Professor of Law and Director of The Law, Science and Technology Program at The George Washington University National Law Center. Mr. Green's extensive experience prompted an appointment as a special adviser to the Director of the National Institutes of Health during the development of NIH's environmental impact statement on the Recombinant DNA Research Guidelines (1976–1978). He also currently serves as counsel for Genex Corporation, a position he has held since the company's inception in 1977.

Mr. Green's article focuses on two broad perceptions that contribute to pressures on corporate and regulatory law in relation to biotechnology: that biotechnology involves special hazards and that this technology has revolutionary implications. Mr. Green considers not only the public perceptions of the environmental hazards but also the moral and ethical issues. Mr. Green examines the present regulatory framework at the Federal level—including the NIH guidelines and related statutes —as well as in the states and local communities. He then examines perceptions of the revolutionary implications of the technology, noting a recent State Department report that classifies genetic engineering, along with unlocking the atom, escaping the earth's gravity, and the computer revolution, as "one of the four major scientific revolutions of this century." Mr. Green compares the evolution of public policy concerning nuclear power to that of biotechnology and concludes by noting that biotechnology is an emerging field whose impact has not yet been tested on corporate law and regulation, but whose impacts could be substantial in the future.

David W. Plant

David W. Plant, the second contributor to this section, is a partner in Fish and Neave, a New York patent law firm. Mr. Plant received his undergraduate and law degrees from Cornell University prior to joining Fish and Neave. He has been a partner in the firm since 1970 and was managing partner from 1981 to 1985. Mr. Plant has handled litigation, arbitration and licensing matters for a number of private companies. He chairs the Committee on Patents of the Association of the Bar of the City of New York. He also served as co-author for a 1982 report on patent protection in the genetic engineering field for that committee.

Mr. Plant's article documents the history of the patent system and the relationships between biotechnology and those historical purposes. He comments especially on patent validity, that many patent law issues have been resolved and that, as a result, numerous biotechnology patent applications have been filed and patents issued. He cites, however, one area of patent law that is still unsettled — the question of genetic engineering with whole plants or animals — and notes special patent requirements in this area for depositing microorganisms with such organizations as the American Type Culture Collection. Mr. Plant next reviews patent infringement, citing special problems in detecting such infringement as well as fairness in compensating inventors of genetically engineered processes and products. Mr. Plant warns that the entire burden of compensation may be borne solely by the first to practice the patented method because the system (after the first use of the patented method) often is self-replicating. He also addresses licensing arrangements among government, universities, and industry and concludes by observing that the patent system is working, but that the key unresolved issue is a method of determining appropriate compensation of innovators. This issue will remain a challenge for all parties involved.

Peter Barton Hutt

The third contributor is Peter Barton Hutt, a partner in the Washington, D.C., law firm of Covington and Burling. A graduate of Yale University, Mr. Hutt received his law degree from Harvard University and his master's in law from New York University, commencing practice with Covington and Burling in 1960 and becoming a partner in 1968. From 1971 to 1975 he served as chief counsel to the US Food and Drug Administration, returning to private practice with his original firm. Mr. Hutt served as a consultant to the Advisory Committee to the Director, NIH, from 1976 to 1978, when the committee held public hearings on the NIH Recombinant DNA Research Guidelines. He joined that committee as a full member in 1978, participating in formulation of many of its health research strategies and serving on its task force devoted to cooperative research relationships with industry.

Using his experiences as reference, Mr. Hutt addresses two aspects of university-industry relationships: the conceptual basis for the relationship and specific patent issues. Mr. Hutt outlines the differing premises on which a university and a company operate — premises strikingly similar to those raised by Donald Kennedy in Part II. Mr. Hutt notes that universities must develop guidelines accommodating the needs of the governmental or industrial group that provides support for the research. Yet that accommodation must not harm the integrity or institutional objectives of the university. Mr. Hutt discusses extensively the matter of academic patent rights for inventions developed under industrial and federal support, the latter governed by the Patent and Trademark Law Act Amendments of 1980. He suggests a number of written guidelines for universities as they deal with industry in new relationships. Since governmental rules generally apply whenever there is a commingling of government funds, Mr. Hutt argues for industry to have the option to segregate its funds from federal funds in academic research, with the corporation bearing the implementation cost. He also discusses the need for rules to ensure that publications or public presentations of research results do not prejudice patent claims in the United States or abroad. Another tough issue for negotiation is the exclusivity of patent licenses. All of these are complex issues, as Mr. Hutt notes, but ones that cannot be avoided. Corporations will seek university relationships, and the public will benefit from these joint ventures in biotechnology. But guidelines and delineated rules must be in place.

The contributors to this section, whose articles originally appeared in *Technology in Society* in 1983 (Vol. 5, No. 2), thus offer an outline of the legal climate for considering the impact of biotechnology on society. Collaborative research relationships among government, universities and industry in this high technology area will also develop from these related legal initiatives. Using the preceding legal, governmental, and academic-industrial discussions as a framework, the articles in Part V will examine biotechnology from international, public, and ethical perspectives.

12

The Impact of Biotechnology on Corporate Law and Regulation

Harold P. Green

The fact that one writes a paper on the "impact of biotechnology on corporate law and regulation" implies that biotechnology has certain unique aspects and note-worthy impacts relating to these subjects. This suggests, as a useful starting point, a discussion of the respects in which biotechnology is unique—or, at least, is perceived to be unique.

At the risk of some oversimplification, there can be identified two broad percep-tions that have contributed to the view that biotechnology has some unique dimen-sions that lead to special impacts on corporate law and regulation. The first of these perceptions is that biotechnology involves special and extraordinary hazards. The second is that biotechnology is seen as involving revolutionary implications. These two perceptions are, of course, interrelated.

Biotechnology—or, at least, the genetic engineering part of biotechnology—has been widely perceived as involving extraordinary hazard to the health and safety of the public and to the environment, both physical and moral. The initial impact of recombinant DNA research and technology dates back to 1974 when a group of the leading scientists in the field called for a moratorium on research activities involving recombinant molecules of DNA because of their view that the technology might involve potentially great hazard and other adverse social consequences. The call for the moratorium was generally respected throughout the world. An international conference at Asilomar, California, in February 1975, called for a continuation of the moratorium on certain experiments and for the establishment of safety guide-lines to govern research in the areas that would be permitted. Such guidelines were promulgated by the National Institutes of Health in 1976.

This "early" history of the technology has had a profound, and probably lasting, influence on its development. It is not surprising that genetic engineering quickly attracted the attention of environmental groups and others concerned about its hazards and social implications. These groups mounted an attack that threatened for a time to rival the intensity of the antinuclear movement. Legislation was introduced in Congress to impose stiff regulatory controls, and, although this legislation did not advance, a number of states (New York and Maryland) and municipalities (*e.g.,* Cambridge, Boston, and Waltham in Massachusetts; Berkeley and Emeryville in California; and Princeton, New Jersey) adopted somewhat restrictive regulations. These state and local regulations generally make compliance with the NIH Guidelines mandatory and establish penalties for violation.

A Revisionist Sentiment

By 1978, however, a revisionist sentiment was in the air. Knowledgeable scientists were, by then, minimizing the risks and were characterizing the former moratorium as an overreaction. Since that time, the NIH Guidelines have been successively relaxed with the inclusion of exemptions from the Guidelines for some types of experiments and the lowering of containment levels for others. This revisionist sentiment does not rest on the proposition that genetic technology is without extraordinary hazard. Rather, it rests on the proposition that prudent practice of the technology and prudent regulation reduce the risk to clearly acceptable dimensions. As eminent scientist Dr. Clifford Grobstein put it in Congressional testimony several years ago, what was perceived in 1975 as "potentially catastrophic and uncontrollable" was by 1977 regarded as only "potentially dangerous but controllable."[1]

A 1981 report by the Congressional Office of Technology Assessment[2] stresses that the risks of recombinant DNA technology are undoubtedly much lower than had originally been feared. On the other hand, the report also notes that "few experts believe that molecular genetic techniques are totally without risk to health and the environment."[3] However, the OTA report was more guarded in its discussion of using genetically engineered microorganisms in the open environment for such purposes as recovery and leaching of minerals, enhanced oil recovery, and pollution control. Since this area of the technology involves large-scale release of microorganisms into the environment, with concomitant lesser control over their behavior and fates, there is a greater likelihood of ecological disruption than is present when the engineered microorganisms are contained within a laboratory setting. The potential risks of use of such microorganisms in the open environment may, according to the OTA report, require government support, including "protection against liability suits."[4] Thus, although, as discussed below,[5] there has been some suggestion that principles of strict liability should apply in the area of recombinant DNA activities, the OTA report suggests the contrary view that practitioners of the technology be shielded from such liability.

Moral and Ethical Hazards

Public discussion about genetic engineering also has reflected concerns that there are also hazards of a moral and ethical nature. These concerns seem generally to be based on the premise that genetic engineering will result in the creation of new forms of life and in human intervention in the processes of evolution. The concerns seem in part to reflect the sense that man cannot improve upon nature, and another sense that in attempting to improve upon nature he is opening a Pandora's box of uncertainty and potential harm. Moreover—even though industrial applications of genetic engineering have nothing to do with manipulation of genetic material in human beings—the concern about such manipulation seems to pervade the entire area of genetic engineering. For example, the OTA report, which deals exclusively with *non-human* applications of applied genetics, regards the social, ethical, and moral concerns about industrial applications as stemming from genetics' "intimate association with people."[6] The belief that genetic technology may lead to the altering of human inheritance, and the relationship in the public's view of genetic engineering to certain debatable aspects of reproductive biology (*e.g.*, sex selection, abortion of defective fetuses, and *in vitro* fertilization) are seen as the bases for the "strong emotions aroused by genetics" even in the purely industrial context.[7]

It seems likely that public concern about the health, environmental, and moral hazards of genetic technology are far from resolved, and that they will recur and arise, with greater or lesser intensity, from time to time in the future. It is important to remember that it is the public's perception of hazard (and not the objective presence of hazard) that is important from the public policy standpoint. Perceptions of risk differ widely and change substantially over time. One need only note the wild oscillations in public perceptions of risk that have characterized atomic energy technology. On several occasions over the past quarter-century, the atomic energy industry took comfort in the "fact" that concern about possible major accidents in nuclear power plants—or on other occasions concern about the adverse effects of discharges of radioactive effluents in normal operation—had been definitively laid to rest; but on each such occasion, the concern again surfaced like Phoenix rising from the ashes.

The fact that genetic engineering's first substantial impact on the public consciousness occurred in connection with the Asilomar Conference in the context of great concern expressed by leading scientists about hazards is a past that is not likely to be lived down quickly or easily. Despite the presently prevailing scientific view that hazards are minimal and manageable, there remains at least a residual, subliminal perception of hazard among the public, and this perception is likely to be invoked and exacerbated from time to time by politicians reacting to the stimuli of events. In this connection, it is not without significance that a number of proposals for Federal and state legislation in the genetic engineering area have been based, at least implicitly, on the premise that these activities involve extraordinary hazard. Similarly, discussions of the issue of liability without fault (which is associated in the legal system with "extraordinary hazard") for injuries resulting from genetic engineering have begun to appear in the law journals, and one legal scholar has described major complexities in potential tort litigation in the genetic engineering area that may require important new liability and compensation mechanisms.[8]

Minimal Risks

The existing pattern of regulation is consistent with the presently prevailing scientific view that the risks of genetic engineering are minimal and manageable. The sole Federal regulatory mechanism to be directed squarely at genetic engineering is the NIH Guidelines which, in fact, are hardly regulatory at all. To begin with, they apply by their terms only to activities funded by the government, and are enforceable only by the sanction of withholding funds. Enterprises that do not engage in such funded activities are subject to the Guidelines only through voluntary compliance, and there is no enforcement mechanism. In addition, as noted above, the Guidelines have been progressively relaxed as the scientific establishment has become more confident that the risks of the technology are quite manageable.

Nevertheless, biotechnology companies and their counsel are acutely aware that the presently benign regulatory framework can change rapidly and dramatically as a consequence of the vagaries of political forces reacting to unforeseeable events. In addition to NIH, other Federal agencies have at least colorable legal authority to regulate biotechnology under such existing statutes as the Federal Insecticide, Fungicide and Rodenticide Act, the Toxic Substances Control Act, the Marine Protection, Research and Sanctuaries Act, the Water Pollution Control Act, the Clean Air Act, the Occupational Safety and Health Act, the Public Health Services Act, and the Food, Drug, and Cosmetic Act. Indeed, the Food and Drug Administration is already actively engaged in reviewing applications for products of genetic engineering, and it is likely that other Federal regulatory agencies will soon be similarly involved. And a really major "flap" could lead to organic regulatory legislation to regulate the technology *per se*.

The President's Commission for the Study of Ethical Problems in Medicine and Biomedical and Behavioral Research in its November 1982 report entitled *Splicing Life* called for a continuing governmental oversight body, "an agency for the protection of the future," which might have regulatory powers as well. Although the Commission's report was directed essentially only to the application of genetic engineering to human beings, at numerous points in the report it was suggested that human applications could not be considered independently of agricultural, industrial, and commercial applications. In Congressional Hearings on the Commission's Report, witnesses expressed some sentiment for establishment of a new "deliberative, non-regulatory" national commission on intervention in human heredity and development to stimulate discussion and understanding of the issues.[9]

There are also intergovernmental dimensions of regulation. It has already been noted that a number of state and local governments have adopted or considered regulatory measures more restrictive than the NIH Guidelines. In 1982, despite the diminution of public concern, both houses of the California legislature enacted such legislation, which was vetoed by Governor Brown. And even in the absence of any real Federal regulatory scheme, some have already suggested that such state or municipal legislation is preempted by the Federal "regulatory scheme," *i.e.*, the NIH Guidelines. In the international context, creation of new life forms has obvious world-wide impacts, raising questions about the desirability of some kind of international regime.

A "Revolutionary" Technology

The second unique dimension of biotechnology is more amorphous, but it can be characterized as the perception of the technology as "revolutionary." Indeed, a panel of experts advising the State Department's Bureau of Oceans and International, Environmental, and Scientific Affairs on the foreign policy implications of the technology recently classified genetic engineering, along with unlocking the atom, escaping the earth's gravity, and the computer revolution, as "one of the four major scientific revolutions of this century." [10] This perception has, of course, contributed to the hazards issues described above, but it has also had other implications from the standpoint of the corporate lawyer.

To begin with, the perceived revolutionary consequences of biotechnology have enormously magnified public reaction to developments in the technology. For example, the first public offerings of stock in genetic engineering companies gained extensive media and public attention. Technical accomplishments are front-page news, at least in the financial press. The fact that "gene-splicing" was involved converted a Supreme Court decision [11] on a prosaic matter of patent law into a burning issue of morality and public policy. Hordes of law students and lawyers, particularly those with prior degrees in the life sciences, are seeking careers in "genetic engineering" areas of the law.

An example of the implications of the perceived revolutionary nature of genetic engineering is the controversy that raged in the late 1970s as to whether the NIH Guidelines and other regulation of recombinant DNA research violated a constitutionally guaranteed right of scientists to conduct research. Although scientists have traditionally asserted that the right to do basic research is constitutionally protected, the constitutional issue involving the Guidelines was spurious and probably would not have been raised but for the perception (primarily by scientists) that somehow recombinant DNA research was different and more important than other areas of scientific research.

While it can be conceded (for the sake of argument) that the right to do research is protected under the Constitution to the same extent as the rights of speech and press, it seems self-evident that the right is far from absolute and can have no greater constitutional dignity than freedom of speech. Therefore, the right can be restricted where there is a rational basis for concern that the research activity might jeopardize health, life, or property. Precedents dealing with the regulation of speech, such as those concerning "fighting words," draft-card burning, and amplified speech, are helpful in drawing the constitutional boundaries of scientific freedom. Surely a scientist has the freedom to think, to do calculations, to write, to speak, and to publish. When, however, the scientist leaves the area of such abstractions and turns to experimentation that may impact on other interests (such as health, safety, or the environment) that are protected by law, he moves within the range of action that may enjoy only some (or perhaps very little or no) constitutional protection. To the extent experimentation could be constitutionally protected, freedom would vary inversely with the degree of perceived impact on persons and on the environment.

Surely scientists — of all people — are, or should be, aware of the fact that scientific research and experimentation have long been subject to prohibitions and regulation without any color of constitutional inhibition. For example, the locus of scientific experimentation may be subject to zoning or other restrictions. No one would argue that a scientist has the right to conduct an experiment by detonating TNT at 3:00 a.m. in mid-Manhattan. And, although some may argue that anti-vivisection laws are unconstitutional, no one would argue that a scientist has the unbridled right to do research on human subjects. Moreover, scientists are well aware of the fact that the government often attaches conditions with respect to the manner that research conducted with Federal funds may be conducted.

The University/Industry Relationship

Another implication of the perception of revolutionary consequences is the new relationship that has emerged between academic and private companies engaged in biotechnology. The commercialization of genetic engineering has involved the necessity for commercial enterprises to acquire the services of scientific talent, the preponderance of which was to be found in academia. Scores of top-ranking biological scientists have been lured from the university classroom and laboratory into the industrial environment on a full-time or part-time basis. To implement the migration of such scientists to industry, biotechnology companies have had to develop new compensation packages and working environments that would be attractive to personnel accustomed to the perquisities of the academic world. In addition, new, complex arrangements have evolved for academic collaboration (*i.e.*, by universities) in commercial and industrial activities.

All of these considerations have had an impact on the world of corporation law and regulation. Although some of these impacts have already had clearly definable effects on legal institutions and relationships, for the most part the impacts are felt in terms of planning for foreseeable future effects.

One is tempted to draw an analogy to the case of atomic energy, another revolutionary development of the 20th century. There are, of course, important distinctions between the two cases, but there are also compelling analogies:

- Scientific experiments and resultant technology are perceived in both cases as having the potential for enormous benefit and catastrophic disaster, including irreversible effects on humans and their environment.
- Painstaking care and stringent regulation can drastically reduce the probability that a catastrophic disaster will occur. Both, therefore, have the capability of a low-probability, high-consequence accident.
- Scientists and technologists create products in both cases that do not exist in nature, and the protection of the public health and safety requires techniques to contain these products in physical structures so as to prevent their entry into the general environment.
- The health, safety, and security of the public rest ultimately in both areas upon faith in the omniscience and infallibility of the human beings who design and implement the scientific and technological endeavors and the safeguard systems within which these endeavors are conducted.
- Both cases raise substantial ethical and moral issues.

Others have called attention, usually pejoratively from the standpoint of genetic engineering, to its similarity to nuclear science. Liebe F. Cavalieri of the Sloan-Kettering Institute has found the similarity to be primarily in the fact that both "confer a power on humans for which they are psychologically and morally unprepared," and he asserts that, although the physicists have already learned this, the molecular biologists have not. He complains that the commercialization of genetic engineering has obscured its "profound ecological, social and ethical implications." His concern is not with immediate hazards to health, life, and the environment, but is directed rather to the inevitability that genetic engineering will deliberately or inadvertently interfere with natural evolutionary processes.[12]

A Benign Technology

There is, however, reason to hope, and perhaps believe, that genetic engineering will not go the way of nuclear science. Nuclear technology was conceived in a military context and its destructive capabilities have been paramount in public discussion. Genetic technology, on the other hand, has emerged in the most benign of all contexts—the fight against disease. Moreover, ionizing radiation is known to have harmful effects on life, but there is no evidence that the engineered microorganisms that will be used in industrial applications will cause harm.

Still, one cannot wholly dismiss the concern that genetic technology may have malign effects. In *U.S.* v. *The Progressive, Inc.*, the case involving the effort of the government to suppress publication of information about the hydrogen bomb, the judge linked genetic technology with the hydrogen bomb in noting that the Federation of American Scientists had suggested that "recombinant DNA could someday surface means of destruction that ought not to be published. . . ."[13]

The major and hopefully decisive difference between the nuclear and genetic engineering technologies is to be found in the attitude of the government and scientific establishments concerned with the technologies. In the case of nuclear energy, the establishment has promised the public a degree of safety, virtually zero risk, that could not be delivered. The over-optimism has been exacerbated for the past 25 years by a pattern of dissembling and propaganda intended to persuade the public that the promise of safety was valid. This has, in turn, created a credibility gap that has detracted from public confidence in the industry and its regulators. The establishment approach to genetic engineering has been quite different. Public confidence has been inspired by a policy and a spirit of full and candid public discussion of the risks, and the government has not assumed the role of sponsor and promoter of commercial and industrial applications. If these attitudes continue, it is unlikely that genetic engineering will face the kinds of attack that have besieged nuclear technology for the past two decades.

To this point in time, therefore, the impact of biotechnology on corporate law and on regulation has been minimal. There have been unique problems, but the uniqueness lies more in degree than kind. The technology is, however, young and immature, and there are distinct possibilities that, at least under certain scenarios, the impacts can be substantial.

Notes

1. Hearings before the Subcommittee on Science, Research and Technology, House Committee on Science and Technology, 95th Congress, First Session, pp. 1043–1046.

2. Office of Technology Assessment, *Impacts of Applied Genetics* (1981), hereafter cited as "OTA Report."

3. *Ibid.,* p. 18.

4. OTA Report, p. 127.

5. See *infra,* note 8.

6. OTA Report, p. 258.

7. *Ibid.,* p. 258–59.

8. Dworkin, "Biocatastrophe and the Law: Legal Aspects of Recombinant DNA Research" in D. Jackson and S. Stich, eds., *The Recombinant DNA Debate* (1979).

9. Testimony of Clifford Grobstein, November 18, 1982, before the Subcommittee on Investigations and Oversight, House Committee on Science and Technology.

10. Department of State, Recommendations of the Genetic Engineering Expert Panel to J. Malone, Bureau of Oceans and International, Environmental and Scientific Affairs, May 15, 1981.

11. *Diamond* v. *Chakrabarty,* 447 U.S. 303 (1980).

12. Testimony before the Subcommittee on Investigations and Oversight, House Committee on Science and Technology, November 18, 1982.

13. 467 F. Supp. 990, 998 (W.D. Wis. 1979).

The Impact of Biotechnology on Patent Law

David W. Plant

This article considers the impact of biotechnology on the United States patent system at points which seem to be among the most sensitive and the least settled.

At the outset, the basic purpose of the United States patent system and the current status of the relationship between biotechnology and the patent system is briefly reviewed. Then the focus turns to issues concerning biotechnology and patent law, which are largely unresolved and which are likely to arise in connection with the enforcement of biotechnology patents through licensing or litigation. These issues relate to patent validity, patent infringement, and the appropriate tribute to be paid to the patent owner.

The Purpose of the United States Patent System

The United States Constitution (Article 1, section 8) provides that Congress shall have the power "to promote the progress of science and useful arts, by securing for limited times to . . . inventors the exclusive rights to their . . . discoveries." In response to that Constitutional mandate, Congress has enacted legislation governing patents (Title 35 of the United States Code).

The first United States patent law was enacted in 1790.[1] The current statute is the Patent Act of 1952, as amended. Federal Courts and the United States Patent Office (now the United States Patent and Trademark Office) have interpreted and elaborated upon the patent statutes. Thus, the Constitution, the patent statutes, and the judicial and administrative interpretations constitute the legal fabric of the US patent system.

A valid and enforceable United States patent grants to the patentee the right to exclude others from making, using, or selling the patented invention in the United States for a period of 17 years from the date the patent issues.[2] A patent does not give the patentee the right to practice the patented invention. The patentee's ability to practice the invention may be limited by a patent to another. Nor does a patent assure income to a patentee. The patentee may lack capital or facilities, or others may simply ignore the patent and corner the market.

In exchange for the right to exclude others from practicing the patented invention, the patentee must describe the invention with sufficient fullness, clarity, and exactness "to enable any person skilled in the art to which it pertains . . . to make and use" the invention.[3] This "enabling disclosure" should appear in the patent specification, which often is comprised of both text and drawings, and which is available to the public when the patent issues.

The underlying theory of the US patent system is that society will benefit from new and useful inventions, and from their prompt disclosure to the public. For example, other workers using the disclosure of a patent can learn from and improve on the patented subject matter. The cost to society of this benefit, and concomitantly the patentee's reward, is the public's lack of freedom for 17 years to practice, without authorization from the patentee, the patented subject matter in the United States.

This underlying theory must always be considered in light of other conflicting policies — the strong policy against economic monopolies (as expressed in antitrust laws) and the strong policy in favor of free access to and use of information and ideas which are in the public domain.[4] Because of these strong policies, patents are critically examined by the Courts in order to prevent patent owners from claiming more than they are entitled to claim. Indeed, in the minds of some, a United States patent affords the owner only the right to *attempt* to exclude others from practicing the patented subject matter. Of course, if that attempt succeeds, damages for past infringement[5] and an injunction as to future infringement[6] may be awarded to the patentee.

Biotechnology and the United States Patent System

Biotechnology and the Patent System Today

Recombinant DNA technology (genetic engineering) and related aspects of biotechnology have matured to the extent that the first commercial products are ready for market (*e.g.,* human insulin, approved by the United States Food and Drug Administration in late 1982). Already, biotechnology and the patent system have had a significant impact on each other.

Many patent law issues relating to biotechnology have been resolved. Numerous patent applications have been filed and patents have been issued in the field.[7] Thus, overarching issues relating to whether a patent should be granted (*i.e.,* whether a genetically engineered, living microorganism is patentable subject matter, whether a deposit of microorganism culture in an appropriate depository satisfies the requirement for enabling disclosure, and the forms of patent claims appropriate to biotechnology inventions) have already been aired in the literature and, to a large extent, resolved by the Patent and Trademark Office and, in a few cases, the courts.[8] But interesting and significant issues remain unresolved.

Categories of Biotechnology Inventions Which May Be the Subject of Patents

As a predicate to this review of some of the unresolved issues, it will be useful to have broadly in mind categories of inventions in biotechnology which may be the subject of patents.

First, in biotechnology, processes have always been regarded as patentable subject matter under 35 U.S.C. § 101, assuming other statutory requisites are met (*e.g.*, novelty — 35 U.S.C. § 102 — and non-obviousness — 35 U.S.C. § 103). Such inventions may include processes for constructing or producing *inter alia* a microorganism, cloning vector, expression control DNA sequence, or structural gene or methods of using such organisms and materials to produce useful products (*e.g.*, desirable polypeptides). One well publicized example of a patent in this class is the Cohen and Boyer patent, United States patent 4,237,224, the claims of which are directed to a method for making recombinant DNA molecules and for producing polypeptides. (The related Cohen and Boyer patent application S.N. 959,288, the claims of which are directed to plasmids incorporating foreign DNA sequences, has not issued.)[9]

Novel genetically engineered microorganisms may also be patentable. In *Diamond* v. *Chakrabarty,* the Supreme Court held that a "live, human-made" microorganism *per se* with "markedly different characteristics from any found in nature and one having the potential for significant utility" is patentable subject matter under 35 U.S.C. § 101.[10] Although the microorganism claimed by Chakrabarty was not the product of recombinant DNA technology, it is generally agreed that *Chakrabarty* is applicable to microorganisms made by recombinant DNA technology.

The merits of the *Chakrabarty* decision have already been canvassed in legal, scientific, and popular literature.[11] For the purposes of this paper, the important point is that the Supreme Court held only that human-made microorganisms fall within the statutory definition of patentable subject matter. The Court did not hold, because the matter was not before the Court, that the particular microorganism at issue met the other statutory requisites for patentability.

It can be inferred from *Application of Bergy*[12] that "pure microorganism cultures" of naturally occurring microorganisms are also statutory subject matter. The Patent and Trademark Office has recently allowed patent claims to such biologically pure cultures.[13]

Because *Chakrabarty* has decided that a living, human-made microorganism constitutes statutory subject matter under Section 101, it follows that components of such microorganisms should also constitute statutory subject matter. Thus, plasmids, phages, and recombinant DNA molecules should qualify as such subject matter.[14] For example, United States patent number 4,273,875 claims "essentially pure plasmid pUC6," as characterized by a restriction endonuclease cleavage map.

DNA sequences may be statutory subject matter, whether produced by recombinant techniques or otherwise.[15] Amino acids, polypeptides, proteins, and enzymes may be patentable.

Genetically engineered, or otherwise "human-made," plant or animal cells can be regarded as statutory subject matter, if an apparently easy extrapolation is made from *Chakrabarty,* which was directed only to microorganisms. However, this extrapolation resurrects all of the philosophical and moral issues briefed in *Chakra-barty* and left undecided by the Supreme Court. Also, human-made cultures of plant or animal cells, regardless of the source of the cell, may arguably be statutory subject matter.[16] For example, United States patent number 4,316,962 claims a cell line established from peripheral blood leukocytes of a 75-year-old man.

That living hybridomas[17] should constitute statutory subject matter seems apparent from *Chakrabarty,* albeit the philosophical and moral issues may be rehearsed here also. Similarly, non-living monoclonal antibodies produced by living hybridomas may also constitute statutory subject matter.

Multi-cellular higher organisms constitute patentable subject matter to a limited extent under the Plant Patent Act of 1930[18] or the Plant Variety Protection Act of 1970.[19] Notwithstanding *Chakrabarty,* however, the question has not been settled as to whether or not (a) plants not within the scope of those two acts or (b) animals may constitute statutory subject matter.[20] This is so, wholly apart from the apparent difficulty in satisfying the Section 112 description requirement as to such organisms.[21]

The net of the arguments and rules on these issues seems to be that, regardless of how organisms, compounds, processes, and products used in or produced by biotechnology are classified, the issues of patentability are conventional except with respect to the living organism itself. As for such organisms, *Chakrabarty* held that a living, human-made *micro*organism constitutes statutory subject matter. But the Courts have not resolved the issue of whether or not eukaryotic cells or multi-cellular higher organisms constitute statutory subject matter. This issue, when raised, can be expected to burn with white-hot intensity.

Alternatives to Patent Protection

An alternative to patent protection is trade secret protection.

In the very early days of biotechnology, some predicted widespread resort to trade secrets in lieu of patents. These predictions have not been realized. Instead, the trend has been to seek patent protection.[22] Indeed, biotechnology's reliance on the expertise of academia is not compatible with the maintenance of trade secret protection. Once a faculty member conducts an open seminar or publishes a paper, trade secret protection may be lost. In such situations, not only patents, but also personal property rights in genetic materials defined by contract, may be asserted in an attempt to protect the fruits of biotechnological research.[23]

As innovation in biotechnology begins to stem more from applied research than pure research, and thus shifts from the academic laboratory to industrial and commercial facilities, trade secrets are likely to assume a more conventional role. Secret processes and recipes, together with unique equipment, are classic subjects of trade secret protection.

Some have urged that copyright protection may be available for the DNA sequences of genetically engineered microorganisms.[24] This theory has not been widely accepted.[25] At present, copyrights are of no more or different value in protecting the fruits of research in biotechnology than in any other field.

The Enablement Requirement of 35 U.S.C. § 112; Deposits

A unique aspect of biotechnology is that, in the case of inventions relating to microorganisms, microorganism cultures are sometimes deposited in authorized depositories in an attempt to comply with the enablement requirement of 35 U.S.C. § 112.[26] This is often so whether the subject matter claimed in the patent is the microorganism itself, a recombinant DNA molecule contained in the microorganism, or a process for producing a product by using the microorganism.

History of Deposit Requirements

The practice of making deposits for patent purposes has developed over a period of more than 30 years.[27] As early as 1949, a deposit was made in conjunction with a patent application on chlortetracycline produced by a microorganism.[28] In 1970, in *Application of Argoudelis,* the practice was recognized by the Court of Customs and Patent Appeals as one way of fulfilling the enabling description requirement of Section 112.[29]

The rationale for requiring microorganism deposits is that, with microorganisms isolated from nature or mutants of such microorganisms, a complete taxonomic description of the microorganism, together with a description of how the microorganism was isolated from nature or what mutation procedures were used to produce the microorganism, might not assure the worker in the art that he would arrive at the same result as the patentee. Thus, such a description might not satisfy Section 112.

The Present Status

Today a microorganism deposit is an accepted way of satisfying the Section 112 enablement requirement for inventions relating to microorganisms.[30] Deposits may be made in an approved depository in either the United States or in a foreign country. The Budapest Treaty governing such deposits, to which the United States is a signatory, requires only one deposit in an approved depository.[31] Thus, in the case of a United States patent application relating to a microorganism, the necessary deposit may be made in an approved depository in a foreign country or in the United States.

The arrangement with the depository should provide, *inter alia,* that the deposit will be available to the public when the United States patent issues. The deposit generally must be made prior to the filing date of the patent application.[32]

Should a Deposit Always Be Made?

The United States patent law does not require a deposit in every case. For example, an invention may be directed to a specific, genetically engineered microorganism. If the untransformed host and unmodified plasmid vector can be described or are otherwise available to the public (*e.g.,* their deposit numbers are known) and if the patent specification describes in sufficient detail the genetic engineering procedures used to produce the microorganism, so that a person skilled in the art, starting with the publicly available untransformed host and unmodified plasmid vector, could accurately reproduce the specific microorganism, a deposit of that microorganism should not be necessary.[33] However, the "ifs" are large, and the safe course is to deposit.[34]

The Section 112 requirements for an enabling disclosure are fundamental to the US patent system. If they are not satisfied by the patent specification, *i.e.,* if the patent specification does not advise the world how to practice the patented invention, the inventor should not be entitled to the right to exclude others for 17 years. Thus, augmenting the written specification with an appropriate deposit may be critically important.

Understandably, private industry has voiced reservations about depositing novel microorganisms, notwithstanding that the depositor may arrange to be informed of requests for the microorganism. The deposited microorganism becomes immediately available to competition when the United States patent issues and possibly earlier (*i.e.,* the date of first publication of an EPC patent application, with some restrictions on access).[35] Having the microorganism, competitors do not need to attempt to reproduce the microorganism using the description in the patent specification or to reverse engineer the microorganism, as is typically the case with respect to patents in other fields. The deposited microorganism gives competitors a ready-made factory which can simply be relied upon to reproduce itself and to produce a desired product, thus potentially saving research expense and start-up time.

However, this, by itself, is no legal justification for failing to make a deposit. A patent applicant must always comply with Section 112. If the conventional written description is not adequate, then the fundamentally important enabling disclosure must be otherwise effected, *e.g.,* by deposit.[36] If it is not, the patent will be invalid.

Patent Infringement and Tribute to the Patentee

The test of patent infringement is set forth in 35 U.S.C. § 271.[37] Pursuant to this section, a United States patent precludes unauthorized practice of the claimed invention in the United States (*supra,* page 2).

Infringement and the Doctrine of Equivalents

The doctrine of equivalents provides that, where there is no literal infringement of a patent claim, infringement may nevertheless be found, if, for example, an unauthorized composition of matter or process employs substantially the same means to achieve substantially the same result in substantially the same way as the subject matter literally claimed.[38]

The doctrine is easy to state. It is often difficult to apply. The field of biotechnology is no exception.

First, the subject matter is complex. Even though microorganisms are among the least complex living things, they are, nonetheless, complicated.[39] As a result of this complexity, a large number of essentially inconsequential variant microorganisms may abound. It is these variants which may have to be reached by the doctrine of equivalents in order for a biotechnology patent to have value to its owner.

It also may be difficult to apply the doctrine of equivalents to inventions claiming DNA sequences and polypeptides. Here, too, variations inconsequential to the invention can exist. Because of the degeneracy of the genetic code, two DNA sequences consisting of different sequences of nucleotide base pairs, can encode for the same polypeptide. Further, most polypeptides contain amino acids which are not essential to the biological function of the polypeptide. As a result, chemically distinct, but functionally equivalent, polypeptides are possible.

These considerations point to the obvious; viz., patent claims must be appropriately drafted so that genetic engineers are not able to avoid infringement by, for example, substituting redundant base pairs or nonessential amino acids.

Detecting and Proving Patent Infringement

Detecting and proving infringement of process claims is frequently difficult in any field. Ironically, one of the effects of the NIH Recombinant DNA Guidelines may be to aggravate this difficulty. A time-honored way to detect infringement has been to look at a suspected infringer's waste. Proper disposal of microorganisms, fermentation media, and the like, as mandated by the Guidelines, calls for their destruction by sterilization. This plainly makes it more difficult to obtain culture samples from those who might otherwise be less fastidious in their disposal practices.

Proper Tribute to the Patent Owner

In biotechnology a potentially vexing issue revolves around the fair basis for compensating an inventor of a method for genetically engineering a microorganism or vector. Once a single microorganism or vector has been produced by the patented method, the method need no longer be used. Thus, the entire burden of compensating the patent owner may be borne solely by the first to practice the patented method. Potentially countless possessors and users of progeny of the original microorganism, even if they are identifiable, may not infringe patent claims directed to the method of genetically engineering the microorganism or vector.[40]

Another aspect of this issue which may frequently be faced in biotechnology is illustrated by recent reports.

In an earlier article in this series,[41] Dr. Richard M. Krause discussed with optimism the prospects for the development of a malaria vaccine:

Because of the biological revolution, there is now the prospect of a malaria vaccine. Had you asked me five years ago about the possibility of such a vaccine, I would have been pessimistic. We had no idea how to develop one. But all that has changed and for the first time there is hope that malaria, and many other of the world's devastating diseases, will be conquered for good.

Unfortunately, the lofty goal of eradicating such diseases may not be so readily attainable. *Science* magazine more recently has reported as follows:

An odd thing happened last fall to New York University's (NYU) project for creating a malaria vaccine. Just as the technical prospects for a workable vaccine brightened, the prospects for getting it manufactured suddenly grew dim. The reason was that one of the lesser funders of NYU's research, the World Health Organization (WHO), could not agree with the proposed manufacturer, Genentech, on how to share the property rights.[42]

Reportedly, Genentech was unwilling to produce the vaccine without an exclusive license and asked for such a license to market the vaccine. WHO contracts provide for public access to the fruits of any WHO-sponsored research. Implementing this provision is clearly not consistent with a manufacturer's desire for exclusivity.

This dispute highlights one special aspect of the-relationship between biotechnology and the patent system. Because some patented inventions in biotechnology have been based on publicly funded research, or have been made in university laboratories, or have a potentially significant impact on the quality of life for large numbers of people, or all three, it may be difficult in such situations to justify the exclusivity normally granted to the patent owner. *Science* has framed one issue in terms of the concern of a hypothetical underdeveloped nation "to find that its citizens' welfare may depend on a business decision made by an obscure company in California. . . ."[43]

Solutions to such concerns may include compulsory licensing or measures such as those contained in recent amendments to the United States patent laws which allow governmentally funded researchers in nonprofit organizations and in small businesses to own patents.[44] Under this new law, as a *quid pro quo* for allowing governmentally funded researchers to own patents on inventions resulting from government-funded research, the United States government is granted a nonexclusive, nontransferable, irrevocable, paid-up license to use the patented inventions worldwide. The statute also allows the United States government to sublicense any foreign government or international organization it wishes, pursuant to treaties or other agreements.[45]

Compulsory licenses are not new in the United States. There is both statutory and case law precedent for them.

Compulsory licenses are provided for in the Clean Air Act (42 U.S.C. § 7608), the Atomic Energy Act (42 U.S.C. § 2183), the Tennessee Valley Act (16 U.S.C. § 831(r)), and the Plant Variety Protection Act (7 U.S.C. §§ 2402, 2404), in cases of compelling public necessity.

Compulsory licenses have also been recognized as a proper form of relief in antitrust actions[46] and in patent infringement actions.[47] In the latter, compulsory licensing has been imposed where an injunction would cause undue harm to the infringer without offsetting benefit to the patentee, or where the possible harm to society by issuance of an injunction against continued infringement has outweighed the benefit to the patentee.[48]

A relevant example is *Vitamin Technologists* v. *Wisconsin Alumni Research F.*[49] The Court ultimately found the patent at issue invalid. But, at the same time, the

Court discussed its power to refuse to enjoin infringement on the ground that the public interest in having access to Vitamin D mandated against an injunction foreclosing the accused infringer from supplying Vitamin D through the sale of irradiated oleomargarine.

Difficult issues also arise with regard to the appropriate basis for compensating biotechnology patent owners pursuant, for example, to licensing agreements. Illustrative is the complex scheme for compensating the owners of the Cohen and Boyer patent.

Stanford University has granted licenses under its Cohen and Boyer United States patent number 4,237,224, and the companion pending United States patent application S.N. 959,288. The license agreements[50] call for initial and minimum annual royalties of $10,000, which can be credited five times against earned royalties. The agreements provide for royalties for the use of the licensed patent rights for the production and sale of marketable goods. Royalties on such "End Products" are calculated on a sliding scale based on annual net sales of marketable goods and range from 1% for net sales up to $5 million to ½% for sales over $10 million. The license agreements provide also for royalties to be paid on products which are not sold or used primarily for further processing or genetic manipulation. These royalties are 10% of net sales of "Basic Genetic Products," 10% of "cost savings and economic benefits" for "Process Improvements Products," and from 3% (for annual net sales of "Bulk Products" up to $5 million) down to 1% (for annual net sales of "Bulk Products" over $10 million).

Tying royalties to net sales of products may be one practicable way of licensing patents on biotechnology processes of producing useful microorganisms, plasmids, etc. It has been previously noted that such a patented invention may only be practiced once in order to arrive at a microorganism, which will, when properly fermented, replicate itself *ad infinitum* without further infringing the patent.

Another scheme to make patented technology available, and at the same time attempt to compensate patent owners, is the recent proposal to pool university-owned patents.[51] The pool would be administered by a clearinghouse, the University Licensing Association for Biotechnology (ULAB). ULAB would issue a blanket license for all patents in the pool, while charging subscribers either a fixed fee or a royalty on net sales. The arrangement is ostensibly for biotechnology patents relating to basic tools of genetic engineering. It would be established to make it easier for universities to license their basic technologies and for industries to acquire them. Notwithstanding its attendant difficulties—the most obvious being potential antitrust problems—the fact that such a system has been proposed confirms that neither academia nor industry should decry the prospect of finding innovative solutions for troublesome issues concerning patents in biotechnology.

The Reverse—The Impact of the Patent System on Biotechnology

The patent system has affected biotechnology dramatically because the spectacular explosion of early innovation in the field so often has occurred in academic laboratories. Fundamental questions as to the relationship of faculty members to industry have been debated.[52] For example, the customary academic practice of promptly publishing has been frequently questioned by those interested in obtaining patents on the early innovations.[53]

Academia's relationships with industrial sponsors in the biotechnology field typically impose restraints on the usual exchanges of information among academic colleagues. Any publication of an invention before a patent application is filed risks forfeiture of patent rights everywhere in the world except the United States, Canada, and the Philippines. Thus, it is not surprising that industry urges delay in publication to allow patent applications to be prepared and filed.

Conclusion

The United States patent system is designed to promote progress in the useful arts. Innovation in biotechnology is occurring at a rapid rate within the ambit of that system. The system clearly can serve, and apparently has served, its intended function by *inter alia* encouraging investors to support innovators in the field, and thus stimulating innovation.

The principal unresolved issue is the determination of appropriate bases for compensating innovators. This is not an easy problem. But it is not the most troublesome problem facing those responsible for dealing with technology in society. Surely it will be addressed and resolved.

Acknowledgment

The author would like to pay special tribute and give grateful thanks to Glenn A. Ousterhout, Esq., and Paul J. Koivuniemi, Esq., for doing the bulk of the work on this paper. I also would like to thank James F. Haley, Jr., Esq., for his helpful critique of the paper. Whatever merit this paper may have is attributable to those three gentlemen.

Notes

1. Law of April 10, 1790, ch. 7, § 1, 1 Stat. 109; 1 Chisum, *Patents* § 1.01 (1978, revised 1982).
2. 35 U.S.C. § 154.
3. 35 U.S.C. § 112.
4. *E.g.*, *Graham* vs. *John Deere Co.*, 383 U.S. 1, 6 (1965) ("Congress may not authorize the issuance of patents whose effects are to remove existent knowledge from the public domain, or to restrict free access to materials already available"); *A & P Tea Co.* v. *Supermarket Corp.*, 340 U.S. 147, 152–53 (1950); the Freedom of Information Act, 5 U.S.C. § 552.
5. 35 U.S.C. § 284.
6. 35 U.S.C. § 283.
7. *Patent Profiles: Biotechnology*, Office of Technology Assessment & Forecast, July 1982.
8. *Infra*, pp. 5–10, 12–16.
9. 1 *Biotechnology L. Rep.* 183–85 (1982).
10. *Diamond* v. *Chakrabarty*, 447 U.S. 303, 305, 310 (1980).
11. *E.g.*, Plant, Ousterhout, and Diana, "Patent Protection for the Fruits of Genetic Engineering," 37 Record A.B. City N.Y. 368 (1982). For a listing of articles, see 1 Chisum, *Patents* § 1.02 [7][d], p. 1–34 n. 24 (1978, revised 1982) (main volume and supplement list 49 articles).
12. *Application of Bergy*, 596 F.2d 952, 975 (CCPA 1979).
13. Halluin, "Patenting the Results of Genetic Engineering Research: An Overview," *Banbury Report 10: Patenting of Life Forms* 67, 82–83 (1982).
14. *Ibid.*, 79–83.

15. Bent, "Patent Protection for DNA Molecules," 64 *J. Pat Off. Soc'y* 60 (1982).

16. *Application of Bergy*, 596 F.2d 952 (CCPA 1979).

17. A hybridoma is a single cell resulting from the fusion of two cells, *i.e.*, a myeloma cell and a lymphocyte. When cultured, a hybridoma may produce useful monoclonal antibodies. Nowinski *et al.*, "Monoclonal Antibodies for Diagnosis of Infectious Diseases in Humans," 219 *Science*, 637–44 (February 11, 1983).

18. The Plant Patent Act is codified at 35 U.S.C. § 161 *et seq.*

19. The Plant Variety Protection Act is codified at 7 U.S.C. § 2321 *et seq.*

20. It has already been urged that the Thirteenth Amendment removes humans from the realm of patentable subject matter. 37 Record A.B. City N.Y. 380 (1982).

21. *Application of Merat*, 519 F.2d 1390 (CCPA 1975).

22. Kiley, "Speculations on Proprietary Rights and Biotechnology," *Banbury Report 10: Patenting of Life Forms* 191–92 (1982).

23. *Hoffman-LaRoche Inc.* v. *Golde et al.*, No. C-80-3601 AJZ (N.D. Cal. filed Sept. 11, 1980); reported in 1 *Biotechnology L. Rep.* 3–6, 12–19, 96, 186–87 (1982). In the wake of *Hoffman*, it appears that both academia and industry are relying increasingly on contractual restraints with respect to distribution and use of genetic materials.

24. Kayton, "Copyright in Living Genetically Engineered Works," 50 *Geo. Wash. L. Rev.* 191 (1982); Kiley, "Learning to Live with the Living Invention," 7 *APLA Q.J.* 220, 230–34 (1979).

25. For criticism of this theory, see, *e.g.*, Cooper, *Biotechnology and the Law* § 11.02 (1982): Whale, "Patents and Genetic Engineering," 14 *Intellectual Property L. Rev.* 93, 110 (1982).

26. Section 112 provides, in pertinent part, that the "specification shall contain a written description of the invention, and of the manner and process of making and using it, in such full, clear, concise, and exact terms as to enable any person skilled in the art to which it pertains, or with which it is most nearly connected, to make and use the same. . . ."

27. The practice was publicly announced in an article by two Patent Office examiners in 1955. Levy and Wendt, "Microbiology and a Standard Format for Infra-red Absorption Spectra in Antibiotic Patent Applications," 37 *J. Pat. Off. Soc'y* 855, 856–59 (1955). In 1959, the Patent Office Board of Appeals rejected an application where no deposit had been made. *Ex parte Kropp*, 143 USPQ 148 (Pat. Off. Bd. App. 1959). In 1963, the Board approved the practice in another decision. *Ex parte Schmidt-Kastner and Hackmann*, 153 USPQ 473, 474 (Pat. Off. Bd. App. 1963).

28. Duggar U.S. patent 2,482,055, issued Sept. 13, 1949. This deposit is mentioned in Cooper, *Biotechnology and the Law* § 5.03[1] at 5–50 (1982), and in Halluin, "Patenting the Results of Genetic Engineering Research: An Overview," *Banbury Report 10: Patenting of Life Forms* 67, 68 (1982).

29. *Application of Argoudelis*, 434 F.2d 1390 (CCPA 1970). See also, *Feldman* v. *Aunstrup*, 517 F.2d 1351 (CCPA 1975) (approving deposit in a foreign depository).

30. Provisions concerning depositing microorganisms are contained in the Manual of Patent Examining Procedure ("MPEP") § 608.01(p). Provisions concerning deposits are also found in the Budapest Treaty on the International Recognition of the Deposit of Microorganisms for the Purposes of Patent Procedure ("Budapest Treaty") (published at 961 *Off. Gaz. Pat. Off.* 21–26 (Aug. 23, 1977)); the Patent Cooperation Treaty Rule 13 bis (published at MPEP § 1823.01); and in the European Patent Convention Rules 28 and 28a.

31. Article 3 of the Budapest Treaty (published at 961 *Off. Gaz. Pat. Off.* 21–26 (Aug. 23, 1977)).

32. Query whether or not a culture from a foreign depository can conveniently clear United States Customs (19 C.F.R. § 12 (1982)).

33. *E.g.*, Kiley, "Learning to Live with the Living Invention," 7 *APLA Q.J.* 220, 228–29 (1979); Kiley, "Patent and Political Shock Waves of the Biological Explosion," 1979 *Pat. L. Ann.* 253, 282; Biggart, "Patentability, Disclosure Requirements, Claiming and Infringement of Microorganism-Related Inventions," *Genetically Engineered Microorganisms & Cells: The Law & Business* 1-1, 2-14, 2-114 to 2-126 (Kayton ed. 1981).

34. Halluin, "Patenting the Results of Genetic Engineering Research: An Overview," *Banbury Report 10: Patenting of Life Forms* 67, 88 (1982).

35. European Patent Convention Rules 28 and 28a.

36. The deposit requirement appears in MPEP § 608.01(p) along with a limitation on the incorporation by reference of external material in a patent application. As for the latter, § 608.01(p) provides that a patent's disclosure cannot incorporate by reference "essential material" unless that material is included in a United States patent or an allowed United States patent application. Clearly, a reference to a deposit, if necessary to comply with Section 112, is the incorporation by reference of "essential material," and is an expressly permitted exception to the general rule set out in § 608.01(p).

37. "§ 271. Infringement of patent

    "(a) Except as otherwise provided in this title, whoever without authority makes, uses or sells any patented invention, within the United States during the term of the patent therefor, infringes the patent.

    "(b) Whoever actively induces infringement of a patent shall be liable as an infringer.

"(c) Whoever sells a component of a patented machine, manufacture, combination or composition, or a material or apparatus for use in practicing a patented process, constituting a material part of the invention, knowing the same to be especially made or especially adapted for use in an infringement of such patent, and not a staple article or commodity of commerce suitable for substantial non-infringing use, shall be liable as a contributory infringer."

38. *Graver Mfg. Co.* v. *Linde Co.*, 339 U.S. 605, 608 (1950):
"The theory on which it is founded is that 'if two devices do the same work in substantially the same way, and accomplish substantially the same result, they are the same, even though they differ in name, form, or shape.'"

39. Dr. James Watson, the Nobel Laureate, has described the complexity of *Escherichia coli*, the microorganism commonly used for recombinant DNA technology:
"Although an *E. coli* cell is about 500 times smaller than an average cell in a higher plant or animal . . . , it nonetheless weighs approximately 2×10^{-12} gram (MW$\sim$$10^{12}$ daltons; a dalton has a MW = 1). This number, which initially may seem very small, is immense on the chemist's scale, since it is 6×10^{10} times greater than the weight of a water molecule (MW = 18). Furthermore, this mass reflects the highly complex arrangement of a large number of different carbon-containing molecules." Watson, *Molecular Biology of the Gene* 65 (1977).
He goes on to add that between 3,000 and 6,000 different types of molecules are present in *E. coli*, ranging in complexity from CO_2 and H_2O to the nucleic acids and proteins. He concludes:
"The structure of a cell will never be understood in the same way as that of water or glucose molecules. Not only will the exact structures of most macromolecules remain unsolved, but their relative locations within cells can only be vaguely known.
"It is thus not surprising that many chemists, after brief periods of enthusiasm for studying 'life,' silently return to the world of pure chemistry." Watson, *supra* at 69.

40. Some of the patent issues in biotechnology are similar to those being confronted with respect to appropriate compensation of copyright holders. In the copyright field, new technology has necessitated new legislation and engendered lawsuits on significant issues, *e.g.*, the Copyright Act as recently amended (17 U.S.C. §§ 101 *et seq.* (1978)); the Sound Recording Amendment of 1971 (17 U.S.C. §§ 101, 102(4)(7), 106(1)(3)(4), 116, 401, 402, 412, 501–504)); *Williams & Wilkins Company* v. *United States*, 487 F.2d 1345 (Ct. Cl. 1973), affirmed by an equally divided court, 420 U.S. 376 (1975) (photocopying); *Fortnightly Corp.* v. *United Artists*, 392 U.S. 390 (1968) and 17 U.S.C. § 111 (1977) (receipt and retransmission of television broadcast signals by cable television); *Home Box Office* v. *Advanced Consumer Technology, Etc.*, 549 F.Supp. 14 (S.D.N.Y. (1981)) (enjoining sale of satellite dishes); *American Television, Etc.* v. *Western Techtronics*, 529 F.Supp. 617 (D.Colo. 1982) (enjoining sale of satellite dishes); and the "Betamax" case, *Universal City Studios* v. *Sony Corp. of America*, 659 F.2d 963 (9 Cir. 1981), 50 U.S.L. Week 3982 (June 15, 1982).

41. R.M. Krause, "Is the Biological Revolution a Match for the Trinity of Despair?" *Technology in Society*, 4(4), 1982, p. 278.

42. 219 *Science* 466 (February 4, 1983).

43. *Ibid.*

44. Public Law 96-517, codified at 35 U.S.C. §§ 200 *et seq.* On February 18, 1983, President Reagan issued a policy memorandum extending P.L. 96-517 to others than small businesses and non-profit organizations:
" 'To the extent permitted by law, agency policy with respect to the disposition of any invention made in the performance of a federally-funded research and development contract, grant or cooperative agreement award shall be the same or substantially the same as applied to small business firms and nonprofit organizations under Chapter 38 of Title 35 of the United States Code.' " (Reported at 25 *Pat. Trademark & Copyright J.* (BNA), p. 351 (Feb. 24, 1983).

45. 35 U.S.C. §§ 200–211, specifically § 202.

46. *United States* v. *Glaxo Group Ltd.*, 410 U.S. 52 (1973).

47. *E.g.*, *Foster* v. *American Machine & Foundry Co.*, 492 F.2d 1317 (2 Cir. 1974), cert. denied 419 U.S. 833 (1974).

48. *City of Milwaukee* v. *Activated Sludge*, 69 F.2d 577 (7 Cir. 1934), cert. denied 293 U.S. 576 (1934).

49. 146 F.2d 941 (9 Cir. 1944 as amended 1945), cert. denied 325 U.S. 876 (1945).

50. Reportedly, 73 organizations have taken licenses under these agreements. 1983 *APLA Bull.* 88.

51. 219 *Science* 1302–03 (March 18, 1983).

52. "Can the Law Reconcile the Interests of the Public, Academe and Industry? (Learning from Experience in Biotechnology)," The Association of the Bar of the City of New York Colloquium (April 21, 1982), summarized at 37 *Record A.B. City N.Y.* 411 (1982).

53. The recent rough sledding in the United States Patent and Trademark Office of the Stanford application directed to plasmids resulted in part from this customary academic practice—specifically from remarks at a Gordon Conference. 1 *Biotechnology L. Rep.* 152–153 (1982).

14

University/Corporate Research Agreements

Peter Barton Hutt

Universities and corporations have engaged in productive cooperative enterprise for decades.[1] It is thus surprising that, with the impressive new advances in biotechnology, the nature and content of these joint enterprises have recently come under public scrutiny.[2] There is nothing inherent in biotechnology that alters traditional issues that have been raised by these joint ventures for decades, nor is there anything inherent in the nature of a university/corporate agreement that differs fundamentally from a contract between two commercial enterprises.

In light of the increased public attention given to these relationships, however, it is appropriate to review both the conceptual premises on which they are based and also the specific patent issues that they must address. This article addresses both aspects of the matter. The author's thesis is that these agreements, properly constructed, can result in substantial mutual advantages for both parties, as well as for the broader public interest, without harming any fundamental academic, corporate, or social values.

Academic Freedom and Corporate Profits

No one can quarrel with the general statements made by university presidents about the essential open and altruistic nature of our great academic centers of learning.[3] A university must indeed exist to foster free inquiry and free exchange of ideas.

Nor can anyone quarrel with the general statements made by corporate officials about the essential profit motive of our great corporate enterprises.[4] A corporation must indeed offer goods or services that will be purchased by the public, at an acceptable rate of profit, or it will shortly disappear.

There is, fortunately, no conflict between these two admittedly different goals. Academic freedom and corporate profits have coexisted and been mutually beneficial for decades. The academic world has educated most corporate officials and has conducted much of the basic research, which has been developed into marketable products by corporations. In turn, corporate profits have been sources of great endowments, yearly gifts, and other funds on which universities depend for their continued existence.

A university must realize that it is one thing to provide an open forum for ideas, but quite another to engage in the enormously expensive type of basic research that characterizes biology and most of the other sciences today. To get from the former to the latter, one needs large sums of money. Since a university cannot conduct basic scientific research without substantial funds, it must forthrightly address the question whence those funds will come and under what limitations.

Because tuitions cannot realistically be raised to cover any significant level of basic research, a university is left with three sources of funding for basic research: the Federal government, private industry, and philanthropy. Of these three potential sources, one would be hard-pressed empirically to argue that any imposes greater restrictions or more onerous limitations than the others. At times, each can be frustratingly parochial and demanding.

Alexis de Tocqueville recognized the problems of financing basic research in *Democracy in America*.[5] He stated that aristocratic societies gave precedence to pure or basic scientific research. In a democracy, however, the profit motive leads inexorably toward emphasis on the application of science in everyday practical terms. He thus argued that "people living in a democratic age are quite certain to bring the industrial side of science to perfection anyhow and that henceforth the whole energy of organized society should be directed to the support of higher studies and the fostering of a passion for pure science."

A Permanent Tension

There is a permanent tension, which will never be overcome, between theoretical research and a market economy. It is the experience of many universities that philanthropical institutions closely resemble industry in their demands for practical applications of science, and Congress is not far behind. As long as Congress is in session, in fact, support for basic research is at peril. Thus, basic research is always in a precarious position.

The pressures on the federal government to increase its limitations and requirements for funding basic scientific research, rather than to reduce them, have existed for many years and are unlikely to abate in an era of reduced economic growth. In the Spring 1978 issue of *Daedalus*, the author catalogued existing public criticism of Federal support of basic scientific research that has come to his attention in the past several years, and urged the importance of dealing forthrightly with the fundamental reasons why the Federal government should fund such research at universities.[6] Unless and until these public criticisms are adequately met, further erosion in public support for such funding could occur.

The author personally supports major government funding of basic research at universities with a minimum of governmental intervention. He firmly believes that de Tocqueville was right. His concern is that this view—which is shared by most scientists—is not sufficiently shared or understood by the public and its elected representatives.[7]

Thus, like most things in life, one cannot reasonably view academic freedom to pursue scientific research as an absolute principle. Any university that insisted on absolute academic freedom, without any limitations or restrictions, would be obligated to decline most governmental, philanthropic, and corporate funding. The only absolute freedom one has is to think. In order to undertake scientific research one must make some accommodation to those who are willing to fund it.

The author does not find this troubling. Indeed, required research to be conducted in the so-called real world of competition and limitations may well be healthier than if true scientific and academic freedom existed in a vacuum. It requires individuals to probe their own motives, to make value choices that are often difficult, to strengthen resistance to the temptation to surrender precious academic principles, and perhaps thus to make better choices and ultimately to produce better research. None of our other most precious freedoms—speech and religion, to name just two—is wholly unfettered.

The real issue, in practical terms, is how far the university should bend in order to obtain research funds, and where it should draw the line. The author would never suggest a wholly pragmatic approach. Some principles must remain paramount. On the basis of substantial experience, however, the author concludes that neither corporate enterprises nor the Federal government has any interest in destroying the concept of academic freedom. Both have an active interest in promoting academic research and an abiding loyalty to their own academic roots and heritages and thus are, in most instances, quite accommodating. The kinds of limitations that are sought by private enterprise are not as inherently in conflict with a university's principles of academic freedom as many in academia fear. If a rule of reason is used on both sides, the relationship should enrich both rather than provoke confrontation.

Academic Competition

In this context, it is necessary to recognize that some of the statements of university presidents about the wholly altruistic nature of universities have been unduly sweeping and pretentious. For those of us whose eyes were first opened by *The Double Helix*,[8] who have witnessed the various cheating scandals in science at major academic institutions,[9] and who have read the public accounts of the race for a Nobel prize between two eminent scientists,[10] those protestations of purity and simplicity do not ring true. While universities can indeed be free marketplaces of ideas, they can also be as motivated by competition as the most cutthroat retailer.

Thus, the contention that commercial developments occurring in recombinant DNA research have drastically altered the tendency towards secrecy in university scientists simply has not been proved. Certainly, Watson and, more recently, Schally and Guillemin would put many corporate entrepreneurs to shame.

It is simply not necessary for either universities or corporations to change their fundamental natures in order to enter into productive research agreements. One need only recognize that universities are not and never have been absolutely open and unrestricted, and that corporations are not and never have been operated solely to make a profit. Where good faith is involved in negotiations between decent people, there is an enormous middle ground on which they can meet.

There are unquestionably compelling reasons counselling against a university directly operating a business. Faculty members must indeed make basic choices in their careers. Once again, however, the author finds it more difficult than many academic administrators to set any bright line between permitted and forbidden faculty involvement with private enterprise.

In Europe, from which the US derives its university tradition, there has always been a very close relationship between private enterprise and the universities. People frequently serve in dual capacities. In the US, eminent scientists like Carl Djerassi have successfully bridged the university/corporate gulf without harming either side.

Nor are many of the current university guidelines on this subject as clear as their drafters intended. Some, for example, would prohibit a professor from serving as director of a corporation, even though that same professor could properly serve as a consultant one day a week — involving an amount of time far in excess of that which would be expended in the capacity of a director, and undoubtedly much greater remuneration. In short, universities run the risk of establishing the same kind of arbitrary rules that academia distrusts so thoroughly when they are imposed by the Federal government and that invite easy evasion.

One must, on the other hand, be fully mindful of concerns about conflicts of interest. Open disclosure of all faculty consulting and research arrangements provides an excellent solution. But if this is a solution for exposing excesses and thus promoting voluntary self-regulation respecting such relationships, why should it not apply equally to directorships and other types of faculty/corporate relationships? Public disclosure in academia can be as important in assuring wide judgment in such relationships as the Freedom of Information Act[11] and the Government in the Sunshine Act[12] have been important in encouraging good government.

Licensing of Patents

Perhaps no aspect of university/corporate agreements has engendered greater discussion than exclusive licensing of patents. The purpose of a university, as all will agree, is to promote the public interest. Experience teaches that, in the area of patents, the public interest can in most instances only be promoted by exclusive licenses. This was the conclusion of the 1968 report of the General Accounting Office (which certainly is not known for ignoring the public interest or giving away government property), which recommends exclusive licenses for NIH patents[13] and, more recently, the Patent Amendments of 1980.[14] In this competitive market economy, private enterprise will not invest funds to develop ideas that can be copied with impunity. Without exclusive licenses, important inventions may remain undeveloped and thus unavailable to the public.

This does not mean, of course, that methods cannot be found to make certain that exclusive licenses are not abused. Reservation of march-in rights can readily serve this purpose.

Similarly, contracting for the exclusive services of a faculty member — where that enables the faculty member to conduct research that otherwise would be impossible — should not automatically be condemned as bringing improper proprietary motives within the university. This implies that money will clearly corrupt faculty members to do things that they otherwise would not do. To reach this conclusion, one must have a very poor opinion of the faculty. A more optimistic approach would be to suggest that, in return for reasonable conditions, faculties be given increased freedom to pursue research of a nature in which they are deeply interested and committed and which may lead to important advances in public health. It is, in short, a matter of how one wishes to view it. University direction and control over research done under an exclusive contract can, of course, be maintained.

It is inevitably the academic humanist, rather than the academic scientist, who most distrusts all university/corporate agreements. Perhaps this is a function of the fact that scientists are far more likely to be approached to enter into such agreements.

Nowhere is the diversity of views on this subject more evident than in the distinction that is often drawn in university policy between faculty rights to copyrighted and patented material. Copyrights (which most often are generated by a university's humanists) are routinely assigned to the authors, whereas patents (which most often are generated by a university's scientists) are routinely assigned to the university. The different handling of these rights is best attributed to historical reasons than to any principled distinction.

Corporate Funding and Patent Issues

Corporate funding of scientific research at a university immediately raises important patent issues. These issues are complicated by the fact that, in most instances, the corporate-funded research is conducted simultaneously with, and perhaps as an integral part of, similar research that is funded by the Federal government. In entering into any university/corporate agreement, therefore, it is essential that the full implications of the arrangement be thought out in advance and reduced to writing.

This requires that both parties have established policies on these matters and are willing to incorporate them into a written contract. While the issues raised by such contracts may be slightly different, in a few instances, from the issues posed by a contract between two commercial enterprises, they are more similar than many people realize and in large part reflect standard contract principles and sound common sense.

Scientific research at a university may at any time lead to an invention that can be the subject of a patent. If that research is funded by the Federal government, the Patent Amendments of 1980 and OMB Circular A-124[15] govern the legal obligations and rights of the inventors and the university respecting any invention. If the research is funded by a corporation, the legal obligations and rights of the inventors and the university respecting any invention are governed by the contract between the university and the corporation.

In the case of scientific research funded in whole or even in the smallest part by Federal funds, the university must comply with the obligations imposed by the Patent Amendments of 1980. As interpreted by OMB, these requirements are not overly burdensome. It is important, however, that these obligations be incorporated in a written university policy in order to fully inform the faculty and students about them. It is also important that they be reflected in any university/corporate research agreement to avoid misunderstanding and confusion about the implications of any related Federal funding.

In the case of a contract between the university and a corporation, the university has greater discretion where no part of the research also depends on Federal funds. Policy must be developed to shape the wise and consistent use of that discretion. It is equally important that a written university policy fully inform the faculty and students about the university's intent respecting such contracts to avoid misunderstanding.

The following issues often arise when inventions are made in the course of scientific research. University policy and university/corporate agreements should provide clearly written rules to resolve them.

Reporting of Inventions

The 1980 Amendments and the OMB Circular contain explicit requirements respecting reporting of inventions by a university to a government agency that is funding research at the university. The obligations of faculty members and students to report inventions to the university (and of the university to report such inventions to the government) should be spelled out in university policy.

Basically the same issues arise under a contract for scientific research between a corporation and a university. The contract should make clear whether the researchers have any obligation to report any invention to the university and should state whether the university, in turn, has any obligation to report such invention to the corporation.

Determination of Inventor

Once an invention is made, the identity of the inventor(s) must be promptly determined. This is often a complex issue in the context of collaborative scientific research. A university policy on this matter should designate a specific office within the university to make the initial determination based upon a review of all the pertinent facts (as well as to carry out other university functions respecting patents). An appeal mechanism should be established within the university, with the ultimate decision resting in a faculty committee or a specific university official. There is no need, however, to include any aspect of this procedure in any contract for scientific research with the Federal government or with a corporation.

Ownership of Inventions

When a student, member of the faculty, or other individual connected with a university makes an invention, the question arises whether the individual(s) or the university, or both, own the rights to that invention. Most university patent policies provide that the university owns or has the right of first refusal to the invention, and that the individual(s) may own the invention only if the university declines to take ownership. Under the 1980 Amendments, the university owns any invention funded by the Federal government (with limited exceptions). When research is funded by a corporation, the contract can provide either that the university or the corporation will own the invention.

The person who owns the invention will be responsible for any patent application. Under the 1980 Amendments, the university must inform the Federal agency about any invention within two months after the inventor discloses it to the university. The university then has 12 months from the date it learned of the invention to determine whether or not to retain title to the invention. If it concludes not to retain title, the Federal agency will obtain title. The university must establish rights to the invention throughout the world or convey those rights to the Federal agency. In order to implement these provisions, the university must require, by written agreement, that its employees disclose all inventions and execute all patent applications and other similar documents as required by the 1980 Amendments and the OMB Circular. This obligation of the university can be discharged through a written university policy.

The same issues arise in a contract with a corporation. University policy and standard provisions should also be developed for such contracts.

Inventions Under Faculty Consulting Arrangements

In virtually all universities there is a written or implied agreement that faculty members may consult with corporations and other groups for specified portions of their time (*e.g.*, 20% or one day per week). Faculty members may also work on inventions on their own time (*i.e.*, at night). University policy should establish what time is the faculty's own time (*e.g.*, weekends and vacations) and whether inventions made outside of university time are subject to the same, or different, rules respecting university/faculty reporting, ownership, and royalties. Any contract governing a faculty consulting arrangement with a corporation should reflect this policy.

Waiver of Patent Rights by the University

Although, as a general rule, the university will usually assert ownership of a patent, in some instances the university may choose to waive the rights to a patent. University policy should specify that, under those circumstances, the university's right to the patent will revert to the inventors. In any contract between a corporation and a university, a specific provision should be included to cover this eventuality.

Separation/Commingling of Private/Government Research Projects and Funds

Under the 1980 Amendments, any invention conceived or reduced to practice utilizing any government funds is automatically subject to the provisions of those Amendments and, thus, to some government rights. Some corporations are willing to permit the commingling of corporate and government research projects and funds, with the result that the 1980 Amendments will apply to any resulting invention. Other corporations may insist upon complete separation of such projects and funds, so that any invention resulting from corporate funding will not be subject to any government rights under the 1980 Amendments.

Any contract between the university and a corporation should deal specifically with the issue of separation/commingling of private and government projects and funds. If the university is unwilling to separate such projects and funding, the contract should so state. If the university is willing to separate such projects and funding, the contract should so provide and specify how it will be accomplished. The contract might also provide, in the event of such separation, that the cost of implementing procedures to assure such separation shall be borne by the corporation.

If separation is provided, the contract should also include a provision respecting the rights of the parties in the event that separation is unsuccessful and the 1980 Amendments become applicable to an invention also funded by the corporation. In all contracts, it would be wise to explicitly recognize that the 1980 Amendments are applicable if government funds are used in the conception or reduction to practice of any invention, in order to avoid the criticism made by the General Accounting Office of the Massachusetts General Hospital/Hoechst contract.[16]

Control of Research

In the course of any scientific research, questions will arise respecting the exercise of scientific judgment about both the general direction and the specific details of the research involved. Which research opportunities to pursue (and how to pursue them) could be a subject of disagreement. This will seldom occur in Federally funded research. It is more likely to occur in research funded by a corporation. The contract between the corporation and the university should, therefore, explicitly state how such matters will be handled.

In most instances, the university will insist upon exercising final judgment on such matters. It may be appropriate, however, to permit the corporation to express its views before the university makes a final decision in order that the university may have the benefit of all possible considerations and to assure a good working relationship.

Trade-Secret Information

The potential receipt of confidential trade-secret information ordinarily should not arise under Federally funded research grants, but could arise pursuant to a contract with a corporation. Some universities are willing to protect the confidentiality of such information. If the university concludes that, as an entity, it has no means of policing the secrecy of confidential information, it should consider including in the contract a provision stating that it will not receive information subject to confidentiality restrictions. Individuals within the university who are working under specific corporate contracts may then enter into separate confidentiality agreements with the corporations to the extents necessary for the conduct of the research. Any special equipment (*e.g.,* a locked file) or procedures could be paid for by the corporation.

Publication and Public Presentation of Research Results

Scientific research inevitably leads to the development of information that the researcher will wish to make public through oral discussion or written publication, but that, if disclosed, may have adverse patent implications. Public disclosure of such information merely triggers the one-year time period for filing of a patent under United States law, and thus poses no significant issue in this country. Under some foreign laws, however, a patent application must be filed before any public disclosure of the invention, and thus such disclosure may preclude foreign patents.

Where the research is funded by the Federal government, the OMB Circular provides that the university must disclose to the government, at the time that it informs the government about the invention, whether a manuscript describing the invention has been submitted for publication and, if so, whether it has been accepted for publication at that time. At any time thereafter, the university must inform the government if any such manuscript is accepted for publication. This allows the government to make certain that foreign patent rights are protected.

In the context of a contract between a university and a corporation, considerably greater discretion is permitted. It would be wise to include explicit provisions respecting not only submission of manuscripts for publication, but also oral disclosure at scheduled symposia or seminars and even spontaneous oral discussions. Reasonable provisions can be found that will protect both academic freedom and the interests of the corporation in preserving patent rights. A copy of any manuscript can be provided to the corporation when it is submitted for publication, which will, in virtually all instances, be several weeks before it actually appears in print. The corporation can be informed about scheduled presentations at seminars or symposia when such presentations are, in fact, scheduled; this, in virtually all instances, will be more than a month in advance. The possibility of spontaneous oral presentations can sometimes be foreseen (*e.g.,* at scheduled seminars or symposia on the general subject matter), in which case the corporation can be forewarned, but if it is not foreseen, the corporation can be informed immediately thereafter so that appropriate steps can be taken to protect patent rights.

No perfect system can ever be established, but reasonable provisions such as these will take care of the vast majority of situations without encroachment on academic freedom or corporate interests.

Exclusivity of Patent Licenses

Under the 1980 Amendments, if the university obtains a patent to an invention funded by the Federal government, the university has the option of granting exclusive or nonexclusive licenses. Ordinarily, potential licensees will be interested in licenses only if they are exclusive for at least minimum periods of time. It was for this reason that Congress explicitly provided that exclusive licenses would be permissible. The 1980 Amendments include specific provisions designed to prevent abuse of an exclusive license, including reservation of a nonexclusive license in the government and government march-in rights.

Where the university conducts research pursuant to a contract with a corporation, the corporation typically will wish to include in the contract a provision stating that any patent resulting from the research will be licensed exclusively to the corporation, subject, of course, to any government limitations and rights resulting from commingling of private and Federal funds. Ordinarily, a corporation will be unwilling to fund research without this type of provision.

The university may conclude that it has a moral obligation to assure itself that any invention resulting from research conducted on its premises and with its resources is used for the public good. Accordingly, various types of mechanisms can be considered to prevent any abuse of such an exclusive agreement. One method is to limit the breadth of the exclusivity to particular types of products or to the line of business of the corporate enterprise involved, thus assuring that any invention could be the subject of noncompeting licenses in other fields that might otherwise go unexploited. This approach raises difficult issues of definition and implementation, however, and the time that must be spent in negotiating such provisions, tailored to each individual contract, may be excessive.

A second approach, which circumvents these problems, is to grant an exclusive license, but to include in it required reports respecting commercialization of the license and a reservation of march-in rights for the university where commercialization is not achieved, but is thought feasible by the university. By incorporating this standard process in each such contract, there is virtually no expenditure of time and effort at the time of the execution of the contract, or until such time as commercialization is found inadequate. The latter approach is, therefore, thought by many to be a more cost-effective way of assuring that the university's obligation to the public interest is fully protected.

Royalty Split Between the Government / Corporation and the University / Inventor

Under the 1980 Amendments, all royalties from a patent obtained from government-funded research are given to the university. There is no recoupment by the government.

In a contract between a university and a corporation for scientific research, the future royalty rights, respecting potential inventions not yet conceived or reduced to practice, are typically resolved. A percentage of sales is usually specified as the royalty for the university. On occasion, the corporation may wish to place a cap on the royalties to be paid in any year or to be paid as a total amount.

Royalty Split Between the Inventor(s) and the University

University policy should determine how the royalties from any invention will be utilized. This is entirely a matter of judgment within the university and should be included in a written university statement of policy, but it is not a subject for inclusion in any contract for scientific research with the government or a corporation.

One or more individuals may be the inventor(s). They may include undergraduate and graduate students as well as faculty and even visiting faculty or people wholly outside the university. A determination must be made whether only faculty, or also students and others, will participate in the royalties, and how those royalties will be divided if there is more than one inventor. It must also be decided whether those royalties will continue to be paid if the individuals should later leave the university and, if so, at the same or at a reduced rate.

The split of the royalties among all the inventors as a group and the university must be decided. The disposition of the university's share of the royalties also must be determined. Some or all of the university's share of the royalties might be designated for future research in the department in which the invention occurred.

Use of Name of University or Researcher

The names of the university and the researcher(s) are valuable assets. Their use by a corporation may enhance the marketability of a product resulting from the research. The contract should prohibit such use or should permit it under specific conditions.

Copyright Rights

In addition to potential patents, the research can lead to potential publications, which are subject to copyright rights. The government does not assert any interest in copyright rights or royalties of this nature. In any contract between a corporation and a university, the copyright ownership and the rights to royalties from any copyrighted materials should be specified.

Complex Issues in a Complex Society

While these may seem like complex issues, they cannot be avoided by any university in today's complex society. The intelligence and productivity of the university faculty make it inevitable that patentable ideas will emerge, and thus that these issues must be confronted. Corporations, recognizing the public value of sound new ideas generated by the faculties of our great universities, will inevitably seek them out. The public will benefit from these joint ventures, because it is only through commerical enterprise that the most important university-generated ideas can be fully utilized for the public good.

Notes

1. *E.g.*, L.E. David and D.J. Kevles, "The National Research Fund: A Case Study in the Industrial Support of Academic Science," *Minerva* 12 (1974), p. 207; National Commission on Research, *Industry and the Universities: Developing Cooperative Relationships in the National Interest* (1980); J.W. Servos, "The Industrial Relations of Science: Chemical Engineering at MIT, 1900–1939," *ISIS* 71 (1980), p. 531; and C.E. Kruytbosch, *Academic-Corporate Research Relationships: Forms, Functions, and Fantasies* (Washington, DC: The National Science Foundation, 1981).

2. For information on such agreements, see "Government and Innovation: University-Industry Relations," hearings before the Subcommittee on Science, Research and Technology of the Committee on Science and Technology, US House of Representatives, 96th Congress, First Session (1979); C.E. Kruytbosch and D.B. Weisz, *Notes on Practices and Policies in Selected Federal Mission R&D Agencies Concerning University-Industry Research Relationships* (Washington, DC: The National Science Foundation, 1980); B.J. Culliton, "Biomedical Research Enters the Marketplace," *New England Journal of Medicine* 304 (May 14, 1981), p. 1195; "Commercialization of Academic Biomedical Research," hearings before the Subcommittee on Investigations and Oversight and the Subcommittee on Science Research and Technology of the Committee on Science and Technology, US House of Representatives, 97th Congress, First Session (1981); B.D. David, "Profit Sharing Between Professors and the University?" *New England Journal of Medicine* 304 (May 14, 1981), p. 1232; W. Lepkowski, "Research Universities Face New Fiscal Realities," *Chemical and Engineering News* (November 23, 1981); D. Kalergis, *The Role of the University in the Commercialization of Biotechnology* (Charlottesville, VA: University of Virginia School of Law, 1981); "The View from the Whitehead Institute, *SIPIscope* 10:2 (March–April 1982); J.L. Teich, *Inventory of University-Industry Research Support Agreements in Biomedical Science and Technology*, National Institutes of Health (January 1982); W. Lepkowski, "Academic Values Tested by MIT's New Center," *Chemical and Engineering News* (March 15, 1982); "University/Industry Cooperation in Biotechnology," hearings before the Subcommittee on Investigations and Oversight and the Subcommittee on Science, Research and Technology of the Committee on Science and Technology, US House of Representatives, 97th Congress, Second Session (1982).

3. *E.g.*, D.C. Bok, "Business and the Academy," *Harvard Magazine*, May–June 1981; A.B. Giamatti, "The University, Industry, and Cooperative Research," *Science* 218 (December 24, 1982), p. 1278.

4. *E.g.*, T.D. Kiley, *Licensing Revenue from Universities – Impediments and Possibilities*, Genentech Corporation, December 15, 1982.

5. A. de Tocqueville, *Democracy in America*, Vol. I, part 1 (1840), ch. 10.

6. P.B. Hutt, "Public Criticism of Health Science Policy," *Daedalus*, Spring 1978.

7. *E.g.*, the three Congressional hearings cited in note 2 *supra*.

8. J. Watson, *The Double Helix* (1968).

9. "Fraud in Biomedical Research," hearings before the Subcommittee on Investigations and Oversight of the Committee on Science and Technology, US House of Representatives, 97th Congress, First Session (1981); and W. Broad and N. Wade, *Betrayers of the Truth* (1983).

10. N. Wade, *The Nobel Duel* (1981).

11. 81 Stat. 54 (1967) and 88 Stat. 1561 (1974), 5 U.S.C. 552.

12. 90 Stat. 1241 (1976), 5 U.S.C. 552b.

13. General Accounting Office, *Problem Areas Affecting Usefulness of Results in Government-Sponsored Research in Medicinal Chemistry*, Rep. no. B164031(2) (August 12, 1968).

14. 94 Stat. 3025 (1980), 35 U.S.C. 200 et seq.

15. *Federal Register* 47 (February 19, 1982), p. 7556, superseding *OMB Bulletin* no. 81-22, *Federal Register* 46 (July 2, 1981), p. 34776, which had provided for interim implementation of the 1980 Amendments. For an extensive analysis and discussion of the impact of the 1980 Amendments and the OMB Circular, see Advisory Committee to the Director of the National Institutes of Health, *Cooperative Research Relationships with Industry* (October 1981).

16. General Accounting Office Letter Report no. B-204687 (October 16, 1981).

Part V

15

Introduction

Joseph G. Perpich

This section builds on the previous chapters, positing issues that governmental institutions, industry, academia, the scientific community, and the public at large must attend to in the 1980s and beyond. Among the topics considered are the international aspects of this technology, the public perception of biotechnology, and the impact of that public perception on governmental and nongovernmental bodies. Finally, these articles, which originally appeared in *Technology in Society* in 1984 (Vol. 6, No. 1), present issues regarding the far-reaching applications of genetic engineering on humans.

Sir Gordon Wolstenholme

The first contributor is Sir Gordon Wolstenholme. Sir Gordon's distinguished career in medical and scientific affairs has included such posts as President of the Royal Society of Medicine from 1975 to 1978 and membership on the General Medical Council from 1973 to 1983. He is also an Honorary Fellow of the Royal Society of Medicine and a Foreign Honorary Member of the American Academy of Arts and Sciences. Sir Gordon served as chairman of the Genetic Manipulation Advisory Group, a body responsible for review of all recombinant DNA research and development of resulting products in the United Kingdom. His was a key role, responsible for ensuring uniformity of guidelines and rules to govern this research in England and for coordinating the group's efforts with the National Institutes of Health and European bodies. It was that close collaboration, particularly between the NIH Director, Donald S. Fredrickson, and Sir Gordon, that led to the fact that NIH and British standards became the model for similar research guidelines throughout the world.

In his article, Sir Gordon surveys events in Britain leading up to the creation of the Genetic Manipulation Advisory Group (GMAG). He observes that the GMAG was a modern British compromise, a constructive "muddle," that aimed at treating each recombinant DNA proposal on its particular merit so case law could develop. He distinguishes his working group from that of the NIH by noting that the British group lacked authority to support outside research through grants. Based on his collaboration with the NIH and international and European scientific associations in developing standards for research, Sir Gordon argues for an advisory group to deal with the vast number of scientific questions that biotechnology poses. One meeting at the Salk Institute discussed a plan by Jacob Bronowski to set up a national panel to assess the wide implications of advances in science on all forms of life and the environment. Sir Gordon notes that the GMAG had the elements of such a model, and he concludes by asking whether small groups, such as national panels similar to the GMAG or a worldwide federation of groups, could ever become a reality for providing oversight on the broad impact of genetic manipulation.

Nicholas Wade

The next contributor is Nicholas Wade, a member of the editorial board of *The New York Times*. Mr. Wade was formerly the deputy editor of *Nature* and a member of the news staff of *Science*. It was in the latter position that he covered closely events involving recombinant DNA technology, beginning with the Asilomar Conference and the ensuing policy debates in Washington involving the NIH guidelines and related federal policies. His years of involvement in reporting on the Recombinant Advisory Committee and the public hearings held on the evolution of the guidelines gave him unique insight into the policy debates in Washington and at the state and local level. In fact, based partly on that work, he wrote *The Ultimate Experiment*.

Mr. Wade opens his article by noting the high regard that the public holds for science and biology, but he points out that the autonomy of science will be a critical issue for biologists in the years ahead. Beginning with the discovery of DNA in 1953 to the invention of gene-splicing in 1973, biologists worked in obscurity. But the Asilomar Conference made the public aware of the science itself, and its potential benefits and hazards. However, Mr. Wade takes biologists to task, for in his view they began then to withdraw from the debate over potential hazards in light of suggested legislation and regulation by federal, state and local governments, and lobbied heavily at the NIH and Congress to relax the guidelines and downplay the hazards. Yet, as Sir Gordon notes, many of the events that Mr. Wade describes can be explained by the joining of the full scientific community with other disciplines in a discussion initially led by the molecular biologists. Most important, microbiologists—who had dealt with hazardous organisms for years—voiced caution and surprise at the concerns raised by the molecular biologists in terms of the potential hazards of this work. Many believed that the fear of potential risks would not be realized. Succeeding years have proved them right.

Mr. Wade's second area of concern regards the commercial aspects of biotechnology, especially in relation to academic scientists. He expresses concern about a number of potential effects, especially in terms of shifting from the focus of fundamental research to applied. However, he notes that, in the end, a system could evolve for the academic scientist in biotechnology no different from that enjoyed by the faculties in physics, chemistry or engineering in their relations with industry. Finally, Mr. Wade concludes that the far-reaching effects of this technology—the application of genetic engineering to human beings—pose ethical problems for society. He notes that the ethical issues have been raised by religious groups, and that the debate will not go away; scientists must join it. Despite all of the concerns he addresses, he concludes that the "admirable sense with which biologists have conducted their affairs hitherto will doubtless prevail in the end."

Alexander Morgan Capron

The next contributor is Alexander Morgan Capron, Topping Professor of Law, Medicine and Public Policy at the University of Southern California. Professor Capron is a graduate of Swarthmore College and Yale Law School, where he served as an editor of the *Yale Law Journal*. After serving as law clerk to Chief Judge David L. Bazelon of the US Court of Appeals for the District of Columbia Circuit, he joined the faculty of the University of Pennsylvania Law School, becoming Professor of Law and Professor of Human Genetics. From 1979 to 1983, Professor Capron was executive director of the President's Commission for the Study of Ethical Problems in Medicine and Biomedical and Behavioral Research, a congressionally chartered commission that reported on a wide range of subjects in the area of bioethics and public policy. The commission's November 1982 report, "Splicing Life — A Report on the Social and Ethical Issues of Genetic Engineering with Human Beings," is the basis for Professor Capron's review and analysis here of the ethical and social implications of this technology.

Professor Capron focuses on the ethical and conceptual uncertainties posed by the application of genetic engineering to human beings. He cites the President's commission statement that ethical uncertainty occurs "when no societal consensus exists as to whether certain applications of gene-splicing are beneficial or undesirable." Conceptual uncertainty "refers to the fundamental change in concepts that this new technology can engender," and Professor Capron describes here the medical advances in diagnosis and treatment using gene-splicing that might raise such complex ethical and social public policy issues. Professor Capron then selects four major areas that illustrate these ethical and conceptual uncertainties.

The effects on human genetic makeup are described in terms of the role this technology might have in undermining the "species barriers" in evolution, namely, the inability of different species to mate and produce fertile offspring. Intergenerational responsibilities, namely, the traditional notions of parental and societal obligations towards children, are challenged by this technology in the detection of genetic abnormalities in the fetus (and the consequential decision whether to continue the pregnancy), and the ability to correct genes that cause diseases and augment other genes to enhance the physical and intellectual capability of the child. These challenges lead to others in the distribution of social benefits — what will be distributed, and who shall receive it — and in the concept of being human — the notion that each individual's characteristics and personality are relatively fixed. These issues surrounding the "improvement" in the human genetic pool have been debated among 20th century geneticists, but advances in genetic technology bring these issues to the fore for general public debate. Noting the need for circumspection in dealing with this technology's potential for unfavorable effects from its very successes, Professor Capron cites Nathaniel Hawthorne's story, "Rappaccini's Daughter," as a more appropriate metaphor than the more frequently used Mary Shelley's *Frankenstein*. However, he concludes that the promise of genetic engineering remains as long as those who practice it are guided by the values and goals of medicine, namely, the application of science and art to human needs.

Lewis Thomas

The final contributor is Dr. Lewis Thomas, Chancellor at the Memorial Sloan-Kettering Cancer Center. Dr. Thomas, a graduate of Princeton University and Harvard Medical School, has held several important academic medical appointments, including the medical school deanships at New York and Yale Universities. In 1973 he became President of Memorial Sloan-Kettering Cancer Center and since 1981 has served as its chancellor.

Dr. Thomas has written extensive essays for a number of prestigious medical journals, including the *New England Journal of Medicine*. Many of his writings have also been collected into books, and one collection, *The Lives of a Cell*, won a National Book Award in 1975. His most recent compilation is *Late Night Thoughts on Listening to Mahler's Ninth Symphony*.

Dr. Thomas's service on a number of government advisory councils includes the Advisory Committee to the Director, NIH, during the periods of public hearings on recombinant DNA research and the later ones on health research strategies to enhance government, university, and industry collaboration.

The essay by Dr. Thomas profiles Professor Oswald Avery, whose work led to genetic engineering and the biological revolution. This article first appeared in *Esquire* magazine's 50th anniversary issue (December 1983), which cited 50 American citizens who had made a significant difference in today's world through individual effort in their chosen field. Of the 50, nine were selected as "trailblazers," having extended our horizons and led us to the "fringe of discoveries." Included among those nine is Dr. Avery.

As Dr. Thomas notes in his essay on Dr. Avery (indeed, as this series of articles conveys), there *is* a biological revolution underway. Here we learn that the roots of those discoveries can be traced to work begun 50 years ago at the Rockefeller Institute for Medical Research by Professor Avery. Avery's work on pneumococcus (a bacterium that causes pneumonia) led to his discovery that genes consisted of deoxyribonucleic acid (DNA), and that DNA was the carrier of the genetic code. His work led to our understanding of genes and their functions. Dr. Richard Krause described in Chapter 5 in this volume how, nine years later, Watson and Crick reported the double-helical structure of DNA, bringing the biological revolution to today's advances.

Dr. Avery's work is an example of the character of scientific research and the intermittent nature of its advances. Dr. Thomas points out that the progress of science is not an orderly succession of logical steps. Progress in basic science, like Avery's, involves a "cascade of surprises." Good scientists must be generally curious, as well as "nervous and jumpy" from ignorance. Such principles have guided the National Institutes of Health in its support of investigator-initiated research grants through a rigorous peer review system that fosters excellence in science, fueling advances such as today's biological revolution.

The authors in this volume address the challenges posed by biotechnology, as well as the enormous promise to meet what Dr. Krause, in his article, called the "trinity of despair"—hunger, disease, and insufficient resources to meet an expanding world population. But we who work with this technology recognize the need for the open atmosphere of inquiry that allows a "cascade of surprises" to occur. And we recognize the necessity of support for basic research from all major partners: government, academia, and industry, and, concomitantly, the indispensable need for circumspection and public oversight in the applications of this technology as addressed by the authors in the concluding series of essays in this volume.

16

Public Confidence in Scientific Research
The British Response to Genetic Engineering
Sir Gordon Wolstenholme

". . . its unique role as an interface between representatives of the public, government, scientists, industry
and trade unions in an area of importance and common
interest."

First Report of the (British)
Genetic Manipulation Advisory
Group (GMAG), 1978[1]

The revelation by American scientists of the imaginable hazards of genetic engineering showed up the national characteristics of the citizens of the United States
and Britain in a predictable and exaggerated form: on the one hand, American
spontaneity, excitability, enterprise, clamor, and open information; on the other,
British hesitation, reticence, caution, passivity, and love of privacy. Although in
the United Kingdom the media tried hard to import their American colleagues'
dramatic exploitation of such alarms as the scientists' "playing God," going against
divine Providence, creating new species, and, more specifically, converting harmless organisms into agents of infection, poison, or cancer, or even giving test-tube
birth to hairy, yellow, green-eyed monsters, the people of Britain remained as scep-
tical and apathetic as ever.

The government and the scientific establishment, not being exposed to public
pressure, came up with typically British proposals for an experiment in compro-

mise. In the opinion of the writer, it proved, nevertheless, to be an experiment of great political significance.

Soon after Professor Berg and his colleagues had called — in July 1974 — for a "moratorium" on experiments in DNA recombination — in order that some assessment could be made of the theoretical hazards which might result from them, the Advisory Board to the Research Councils in the United Kingdom set up a Working Party on "The Experimental Manipulation of the Genetic Composition of Microorganisms" under the chairmanship of Lord Ashby. Within a few months the Working Party produced a report,[2] which was written with exemplary clarity and brevity, and which essentially made the following recommendations:

- only those people with experience in handling dangerous organisms should undertake the experiments;
- the equipment used should be similar;
- granting agencies should inquire — in relation to new research proposals — whether they involved biological hazards;
- biological safety officers should be trained and designated to advise on appropriate procedures;
- the vectors of recombinant DNA should be those not likely to infect man or animals, or those which require nutrients not found in the human gut or which could not survive at body temperature;
- large-scale work involving the use of ten litres or more should be carried out only in specially equipped laboratories;
- tests of pathogenicity should be made on animals, and health monitoring should be instituted for all persons involved in the experiments; long-term epidemiological studies should be conducted of the health records of the investigators and their families; and
- a widely publicized advisory service, perhaps offered by public health laboratories, would help "to safeguard the interests of the public and of those engaged in the experiments."

Given such safeguards, vigorously implemented, the Working Party concluded that the techniques of genetic manipulation should be actively encouraged "because of the great benefits to which they may lead."

Ashby recognized that the techniques would be used in many countries in Europe and elsewhere, and expressed the hope that international agreement, perhaps through the International Council of Scientific Unions (ICSU), would soon be achieved on the conditions under which genetic manipulation of micro-organisms would be carried out.

The Second Working Party

The British government accepted the Ashby report immediately, and proceeded to set up a second Working Party, this time under the chairmanship of Sir Robert Williams, "to draft a central code of practice and to make recommendations for the establishment of a central advisory service. . . ." The Williams Report[3] on *The Prac-*

tice of Genetic Manipulation was published with commendable speed by August 1976. It recommended four levels of containment, as in the United States, designated Categories 1–4, but not quite equivalent to the American P1–P4. Great reliance was to be placed upon local genetic manipulation safety committees, properly representative of all people connected with the experiments — not only scientists and technicians, but also laboratory cleaning and maintenance staff.

In view of the "present state of knowledge of the field, containment measures should allow a suitable margin of safety until any areas of doubt can be clarified by further experimental evidence." There was a "need for a flexible approach . . . by requiring those who plan to work in the field of genetic manipulation to submit their experimental protocols to a central advisory group." In the United Kingdom, this advisory body — the Genetic Manipulation Advisory Group (GMAG) — would be in a position to advise on all such research, and on the development and use of any resulting products, whether in universities, governmental or private institutions, or industrial laboratories.

The scale of research in Britain is, of course, altogether different from that in the United States. The sheer size of the American continent, the multitude of institutions, and the independence and individuality of each state confound any grand Federal design in the realm of science. However, it was a practical proposition in the United Kingdom to establish a dialogue between GMAG and all the people in some 30 to 50 centers where genetic engineering research was to be anticipated. GMAG was a moderate, British compromise, a unique, imprecise, but constructive "muddle," treating every experimental proposal on its merits within the general terms of the Williams Report, and allowing case-law to accumulate. Further, British people are used to having their lives ruled by custom, rather than by statute, rule or regulation.

Details of research projects were to be submitted simultaneously to GMAG and to the governmental Health and Safety Executive (HSE). The latter had in reserve the legal sanctions which could be imposed when workers or the public were considered to be at risk. It proved sufficient for such power to remain in reserve. The notifications to GMAG were at first on a voluntary basis, and were observed in such a responsible and cooperative manner that compulsion — when it came in the form of a government regulation in August 1978 — threatened to be counterproductive.

The Williams Report

The Williams Report was published in August 1976. As with the Ashby Report earlier, the British government accepted the findings without delay, and the Department of Education and Science (DES) began immediately to set up GMAG. The writer agreed to be its first chairman, and was consulted on the composition of the group. Williams had recognized that GMAG would have to command the respect of the public as well as of the scientific community; it needed to include reputable scientists with expert knowledge of the techniques in question and of the relevant safety precautions, and also individuals able to take account of the interests of employees, employers and the general public. We determined at once that we had to have scientists of sufficiently varied expertise to be able to assess all the scientific implications of the projects submitted to us, and—by site visits—capable of judging both the physical conditions in laboratories and the quality of scientific experience and discipline available for the research proposed; able also to serve expertly on subcommittees dealing with such essential matters as the safety of vectors, the adequacy of laboratory equipment, the competence of safety committees, and the conceivable health hazards; and with epidemiological follow-up of causes of morbidity and mortality among those doing the experiments.

The group had to be small enough to have a chance of allowing strong personal relationships to be established. Eight members were selected after consultation with the Confederation of British Industry (CBI), the Trades Union Congress (TUC), and the Committee of Vice-Chancellors and Principals (CVCP)—two to represent management, four on behalf of employees, and four to represent public interest. Many members qualified for inclusion in more than one capacity. As chairman, despite a medical and scientific background, the writer could be regarded as a representative of the public. The management representative nominated by CVCP was a national authority on radiation hazards; three GMAG members also served in the Dangerous Pathogens Advisory Group (DPAG); one member was a general medical practitioner, one a psychiatrist, one an eminent science journalist, another a philosopher and ethicist, while the trade union members included two scientists and a doctor experienced in occupational medicine. It must be emphasized that GMAG was concerned with *Health* and *Safety,* of humans, animals and plants. It could never be competent to assess the scientific merits of a proposal—although one or two lay members tried to extend its duties in that direction—nor could it act as a grant-giving or grant-supporting agency.

We were joined at our meetings by assessors from government departments concerned with Health, Education and Science, Agriculture and Fisheries, the Scottish Office, and the Health and Safety Executive. They remained silent unless called upon by the chairman for an opinion or advice, and took no part in making decisions.

"Exciting, Exhausting and Encouraging"

A skilled secretariat was provided by the Medical Research Council (MRC) on the basis of primary responsibility to the chairman and DES, rather than to the MRC. Riding this unprecedented camel proved to be exciting, exhausting and encouraging. The prospects at first were dismaying. The TUC tried to stick out for increased representation; they were allowed to gain one member, but immediately demanded another, although unwilling to give a place to one of the most relevant unions, which apparently was not in good relationship with the TUC. Some of the lay members started with the deepest suspicion of science, scientists and industry, especially of multinational companies. The scientists were hardly accustomed to taking each other into full confidence, never mind unknown members of the lay public. Industries deeply distrusted the group's reliability to hold in confidence the details of their research projects and potentially patentable products into which they were very cautiously investing considerable sums of money.

Those members of GMAG who felt a responsibility to report back to nominating organizations objected to taking oaths of secrecy—as did one member on religious grounds. Scientists both in universities and in industry took reluctantly to the idea that their proposals had to be approved, and the work monitored by the odd mixture of people who served on their safety committees; naturally, they had no love for additional paperwork, and disliked having to complete forms—in duplicate—detailing their experimental plans and physical and biological containment proposals; still more, they resented the fact that—once given GMAG approval—the procedures could not be varied in any substantial particular without reference back to GMAG—a frustrating delay which was minimized as much as possible by telephone communication. Doctors foresaw incompatibility with the confidences of the doctor-patient relationship, the maintenance of public records, and the protection of the workers, their families, and the public. Environmentalists expected the worst—along lines similar to nuclear catastrophes—and communities regarded the construction of high containment facilities as no more acceptable in a neighborhood than nuclear power stations, airports, or high-security prisons; the first proposal was always to isolate it in somebody else's rural area, despite the absence of housing or sources of labor. Problems ran right down to the young man who—when asked to wear a face mask for his own protection—refused to sacrifice his beard, and to some senior scientists who had "always used a mouth pipette and never come to any harm."

"Fears, Suspicions, and Ambitions"

The inevitable differences of so disparate a membership were magnified by consciousness of the fears, suspicions, and ambitions of the "constituencies they personified." Representatives of the general public might perhaps rely on a continuation of public disinterest, but the scientific members were given frequent reminders that their colleagues throughout the country were acutely sensitive to any infringement of the freedom of scientific inquiry, and were opposed to any restraint on the free expression of their skills and imagination. Scientists had good reason to be worried; the trade union members were aware that they had colleagues outside who saw in GMAG a chance to begin to establish worker control of scientific research in government institutions, universities and major industries; at the same time the unionists realized that too restrictive an attitude might result in the loss or absence of jobs for their scientific and technical members.

During the meetings of GMAG a single injudicious remark was likely to produce instant polarization, fatal to the desired consensus. Usually such crystallization could be dissolved by humor; only once were we driven to take a vote—on confidentiality, when those who refused to be bound to secrecy were subsequently barred from consideration of "sensitive" proposals.

Fortunately, the chief characteristics of the membership proved to be good humor, intelligence, patience, and faith in the ultimate benefits. No praise could be too high for the manner in which our scientists made the principles of their esoteric techniques painstakingly comprehensible, even to those with absolutely no education in science; equally, the lay members patiently strove to understand the points that mattered in regard to risk, safety and health. Every member found an important aspect of the procedures in which they could take an informed interest on behalf of the group, such as the proper composition of safety committees, the safe transport of research materials between national and international centers, the confidentiality of medical records, the psychological stress of working in high containment conditions, the design of protective cabinets, the special facilities required for work with plant viruses, or the medical contra-indications for work on genetically altered organisms. It was teamwork at high tension, but also at a high level of mutual trust and, ultimately, of friendship.

The group began its work at the end of 1976. In the first year the British decisions on categorization of physical and biological containment proved in practice to be less restrictive than the early *Guidelines* of the National Institutes of Health (NIH). As a result, most other countries in Europe were inclined to follow the GMAG example, and more of them might have done so if GMAG had been in a position to publish the full details of the proposals and their categorization. Then American scientists began profoundly to regret the clamorous response which their responsible expression of anxiety had stirred up. They put pressure on the NIH to modify and relax the containment regulations; microbiologists, in particular, drew public attention to the tried safety of work with dangerous organisms. As the United States relaxed its rules, the British GMAG — at that time intent on experimental testing of the risks involved, if any, was slow to follow, and one result of this was to make the revised NIH Guidelines widely preferable in Europe to the British rules. So long as international agreement was extended, it did not matter which system was adopted; the one danger was that a country without adequate regulations might prove an experimental haven for unscrupulous exploiters of these new, richly promising techniques. Fears were also often expressed of secret developments for military purposes.

GMAG in the International Context

GMAG's work had its international context. Through the personal kindness of the then director of NIH, the chairman was kept informed of the ups and downs of American opinion and policy. GMAG joined in the deliberations on recombinant DNA experimentation of the European Economic Community (EEC), the European Science Foundation (ESF), and the European Molecular Biology Organization (EMBO); and we formed links with the International Council of Scientific Unions (ICSU) and its committee on genetic experimentation (COGENE), and with the World Health Organization (WHO) in regard to safety measures in microbiology and the international transfer of research material. Friendly exchanges took place with bodies equivalent to GMAG, which were attached to research councils in other countries, notably Canada, Australia, Sweden and Japan.

A sadly instructive lesson was learned in relation to the EEC. At the beginning of GMAG's work, the chairman took part in discussions in Brussels between representatives of the Commission and scientists involved in the regulation of genetic engineering research in member states of the Community. He was personally much in favor of an immediate legally enforceable Directive, creating a general framework on British lines within which the research could go ahead confidently on an agreed basis throughout most of Europe.

The British government, however, led the way in opposition to any Community ruling, on the grounds that the Treaty of Rome (which established the EEC) could not be applied to scientific research. When, three years later, the EEC at last tried to introduce a Directive with the consent of the governments, the writer felt strongly that the good sense of scientists, who were behaving with full responsibility, would be affronted by this late exercise in compulsion. Supporters of legal sanctions in Europe included those unionists who still hoped to use genetic engineering to gain acceptance of political intervention in science. With difficulty, and after much argument, the EEC agreed to step down its Directive to the level of a Recommendation, which has a moral, guiding influence, but no legal standing.

The wide acceptance of the American view that most genetic manipulation experiments posed no risk to man, animals, or plants, beyond those normally encountered in microbiological research, had the effect, as stated earlier, of eclipsing the more cautious British attitude in Europe and elsewhere. The consequent downgrading of GMAG into a technical advisory committee of HSE need not, perhaps, be regretted, but it is having one very unfortunate result. The memory is rapidly fading of an important experiment in communication between society and science. Years ago, the writer was asked to visit the Salk Institute in La Jolla to discuss a plan by Bronowski to set up a national panel of "wise men" to assist the American public to assess the implications of advances in science for all forms of life and for the environment. One extra, small voice could do nothing to help make this a reality, not even on a trial basis. Perhaps for well over 200 million people the idea is entirely impractical. But for around 50 million — in the United Kingdom — GMAG demonstrated that something of the kind is not impossible.

The Current and Imminent Problems

The public needs more than the media can or do provide, especially when we consider some of the current and imminent problems, which are capable of arousing serious anxiety, for example, accumulation of radiation exposures; disposal of toxic wastes; acid rain; environmental induction of cancer; abuse of drugs; inability to distribute subsistence foods; anachronistic inequalities in the enjoyment of health; scarce mineral resources; mass unemployment; illiteracy; excessive nationalism and tribalism; religious fanaticism; and, in relation to genetic engineering, prenatal determination of human genetic constitution; correction or modification of inherited defects; intervention in brain capacity and function; preservation of deep-frozen human embroyos; and the creation of strange, interspecific forms of plants and of animals.

So much for many of the fears. The public is equally entitled to receive advice in an informed, balanced way about contributions which genetic manipulation (and other scientific advances) may be expected to make towards the solution of some of the world's most pressing problems: new foods, grown where most needed, disease resistant and independent of costly and ecologically harmful fertilizers and pesticides; new sources of energy; conversion of wastes; many more and safer vaccines to protect against infections and infestations; anti-viral and anti-cancer agents; cheaper and readily available hormones and drugs for human and veterinary medicine; enzymes for a multitude of valuable purposes; and, no doubt, discoveries beyond our present comprehension.

Our nations and our peoples deserve the advice of people who are prepared to devote their intelligence, knowledge, and humanity to the deepest possible consideration of such eventualities. An advisory team of this kind would need to become practiced and experienced in such enquiries and debates over a longish period. Only small committees are effective, but a national panel of the kind proposed should probably include one representative, at least, of government, science, medicine, law, education, religion, ethics, ecology, ethology, the trade unions, and industry.

GMAG may have been a small beginning, in one country. But if we do not look beyond our own nations, we are surely doomed. Could such a group—or federation of groups—on a world, planetary basis, ever become a practicality?

Notes

1. Cmnd. no. 7215 (London: Her Majesty's Stationery Office, 1978).
2. Cmnd. no. 5880 (London: Her Majesty's Stationery Office, 1975).
3. Cmnd. no. 6600 (London: Her Majesty's Stationery Office, 1976).

17

Biotechnology and Its Public

Nicholas Wade

Biotechnology has opened up large cracks in the high wall that shields the laboratory from public gaze. The public already understands that, in genetic engineering, today's research discovery may be tomorrow's new product. Biologists' conduct of their affairs will probably come under closer public scrutiny than any group of researchers has faced hitherto. In meeting that test, they have both assets and handicaps.

A vivid indication of public interest in biology was the statement issued in June 1983 by a wide spectrum of Protestant, Catholic and Jewish leaders. The group called for a ban on making inheritable changes to the human genome. Since its members represented the heads of most major denominations in the United States, the statement served at the least as a warning of the kind of debate that biotechnology is likely to engender.

Biology's Public Standing

Despite quite frequent assertions to the contrary by some scientific spokesmen, the public holds science in high regard. The biennial surveys conducted for the National Science Board and published in its Science Indicators series consistently portray strongly favorable attitudes toward science, and demonstrate the public's ability to distinguish between science and the adverse effects of certain technologies.

Biologists capitalized on this goodwill during the late 1970s by themselves raising and addressing the issue of recombinant DNA hazards. Provided that at least a figleaf of the original safety rules is kept in place, that self-denying posture will serve to argue that biologists can continue to be trusted to run their own affairs.

The question of the autonomy of science will be a critical issue for biologists and the public in the years ahead. For exactly 20 years, from the discovery of the structure of DNA in 1953 to the invention of gene-splicing in 1973, molecular biologists enjoyed the luxury of working in decent obscurity. During that long and fecund gestation, the groundwork of a powerful discipline was laid. A large body of theory was established about the molecular biology of viruses and bacteria, even if eukaryotes continued to remain intractable. New and more powerful techniques were devised. Besides the recombinant DNA technique itself, there were the two methods of rapid DNA sequencing developed by Sanger and by Maxam and Gilbert. Köhler, working in Milstein's laboratory, invented the monoclonal antibody technique in 1975.

The citizens of the United States and Europe, benevolent patrons who paid for these good works, had given rather little attention to what was being done with their money. When their attention was drawn to molecular biology by the debate over the hazards of gene-splicing, they viewed the new science as though seeing it for the first time. Like Athena springing from the head of Zeus, molecular biology was born fully grown. The public's first impressions were favorable. Here were the midwives of the new science already debating in advance of any danger how a powerful new tool could best be wielded so as to do the public no harm.

The debate burst into public view with the convening of the Asilomar conference of 1975. What prompted the conference was a chain of events that began when Robert Pollack, then of Cold Spring Harbor, learned of an experiment in which it was planned to insert SV-40 virus into Escherichia coli, the leading denizen of the human gut. Some biologists with no sense of history have argued that the experiment was always clearly harmless, and that, if only they had been consulted at the time, they would have said so. In fact, the hazard posed by the experiment was so novel, and crossed the demarcation lines of so many different disciplines, that no single expert in 1971 was qualified to answer Pollack's objection.

The Asilomar Conference

The Asilomar conference laid the basis for addressing that issue, and biologists deserve enduring credit for denying themselves full use of the new technique that everyone was impatient to use. But a few spots began to appear on the stainless record after 1976, when the National Institutes of Health issued its recombinant DNA safety rules.

Up to that point, the possible hazards of the technique had been freely discussed among scientists. But when Congress and local communities began their own discussions of whether the safety rules were adequate, as they had a perfect right to do, the tenor of the debate changed abruptly.

The dissent from within the scientific community evaporated overnight. An opinion poll taken by *The Boston Globe* showed that many biologists seriously doubted that the NIH safety rules were stringent enough, and yet, with honorable exceptions such as Ernest Chargaff and Robert Sinsheimer, few voiced their concern in public. The young researchers who had vocally protested the laxity of the earlier versions of the safety rules fell suddenly silent. They claimed, whether rightly or wrongly, to be under great pressure from their elders to say nothing to rock the boat or entice politicians to bring research under tighter control.

Free and open debate is the heart of science. What prompted biologists to abandon this central value was fear: fear that the public would fail to understand the hazard issue, would demand absolute safety and, worst of all, might intrude an alien presence into the laboratory itself, demanding a veto over every proposed experiment.

Events proved that fear to be misplaced. Both in Congress and at the local level, as with the Cambridge citizens council, the public proved to be both understanding of the biologists' concerns and tolerant in its recommendations. Throughout the political debate, scientific lobbyists kept protesting that the public didn't seem to trust scientists. The truth was just the opposite: It was scientists who didn't trust the public.

True, Congress played for a time with the outrageous notion of writing a law that scientists should abide by the safety rules that they themselves had written. This gross affront to the dignity of the scientific profession elicited vehement reflexes. The flag of Galileo was waved *con brio*. Congress was accused of embarking on a Lysenkoist crusade to hobble genetics and stamp out freedom of inquiry. Opponents were castigated for the extremism of their arguments.

This show of affronted dignity was somewhat undercut by the discovery that biologists at two leading institutions, the University of California, San Francisco, and the Harvard Medical School, had indeed been unable to abide by the safety rules. The infractions, though minor, were highly inopportune. Biologists at UCSF, told by NIH that a certain plasmid was not certified for use, went ahead and used it anyway. Entries in the lab's logbook were falsely recorded. When the incident came to the attention of the department's biohazard committee, a lukewarm investigation cited the NIH as chief culprit. The infraction at the Harvard Medical School was committed by a biologist who, as a member of the NIH recombinant DNA committee, had himself helped to draw up the regulations.

Departures from the Scientific Ethos

Had Congress held an ounce of malice toward science or scientists, these manifestations of indiscipline would have furnished ample pretext for imposing draconian rules on research and a roster of fines and forfeits. In fact, Congress's reaction was astonishingly mild, and in the end it was content merely to exercise oversight without passing any legislation.

Biologists, of course, had every right to lobby against legislation they didn't like. But there were disconcerting departures from the scientific ethos in some of the measures used in the lobbying campaign. Experiments with little real relevance to safety were announced to the press before publication and their significance was grossly overplayed. Political lobbyists routinely use information this way; couldn't scientists, even in the heat of the moment, have held themselves to a slightly higher standard?

Similar methods were in evidence as biologists put pressure on the NIH to reduce the stringency of the safety rules. Conferences would be held at which interested experts would pooh-pooh the possible risks of recombinant DNA. Typically there would be little new evidence beyond that which the NIH Recombinant DNA Committee had already considered in drawing up its rules. But the "consensus" of experts would be cited as proof that the rules should be rolled back in this area or that. The conclusions of one of these meetings, the Falmouth conference, notes Sheldon Krimsky in his book, *Genetic Alchemy,* "were extended far beyond their original intent. Even the highly restricted scope of these conclusions . . . is not based on extensive data or rigorous argument; it was reached on the plausibility of the evidence."

But these were small blemishes on the shining escutcheon of Asilomar. With or without legislation, with tight or relaxed safety rules, biologists had established so responsible a reputation that the next phase of their interaction with society, the commercialization of molecular biology, attracted rather less scrutiny than it probably deserved.

The Commercialization of Molecular Biology

Until recently, molecular biologists disdained reaping financial rewards from pure research. Cohen and Boyer at first neglected to patent the recombinant DNA technique they invented; when urged by the Stanford University patent officer to do so after he had read an account of the invention in *The New York Times,* the two reluctant inventors agreed to apply for a patent only on the condition that any royalties be turned over to their respective institutions. Some argue this attitude was naive; chemists and engineers have been consulting for industry for years. Nonetheless, it represented a widely held attitude among biologists at the time.

Since then, there has been an almost complete reversal of views. Biologists *en masse* have formed commercial ties with industry or set up their own companies. What is disturbing about this development is its scale. "It is already apparently true that there is no notable biologist in this field anywhere in America who is not working in some way for a business. I interviewed some two dozen of the best molecular biologists in the country and found none," wrote *Washington Post* reporter Philip Hilts in 1982.

The heavy involvement of biologists in commercial applications deprives the public, at least for the moment and maybe longer, of an independent source of advice. Certainly, if the DNA hazards debate were to be replayed, or resumed, biologists who denied the risk would be accused of self-interest and their arguments discounted. Beyond that, the widespread commercial interests of all leading practitioners cannot fail — despite their individual denials — to have some impact on the overall direction of research, probably shifting it toward practical goals at the expense of knowledge for its own sake.

Engineers, of course, see nothing wrong with that; in their discipline many advances have come from tackling applied problems. But biology is not necessarily the same. Not only may the direction of research be shifted, but also the tradition. Harriet Zuckerman's study of Nobel Prize winners brings to light how many Nobelists received their training at the hands of other Nobelists. What kind of research tradition will be inculcated into today's crop of graduate students by professors who are all engaged in pursuing money as well as truth?

Perhaps the present high degree of commercialization of academic biology is merely a temporary phase. Perhaps the applied side of molecular biology will become sufficiently divorced from pure research that the same individuals will not be in such heavy demand for both activities. More likely, biologists, having once tasted riches, will not retreat to the cloister or resume their vows of poverty.

A Third Phase of Debate

What will that do to their public image? After hazards and commercialization, a third phase of debate has just begun, that of the ethics of biotechnology. Biologists have not so far established particularly striking credentials in this arena. The first attempt to apply genetic therapy to humans was undertaken by Martin Cline of the University of California, Los Angeles, who inserted human globin genes into the marrow cells of patients suffering from beta-thalassemia. There was a touch of envy in the criticisms of some of Cline's colleagues that the experiment was premature; it was in some ways a bold initiative that, in different circumstances, might have drawn praise. Unfortunately, Cline failed to inform fully some five oversight committees as to his intention of performing the experiment. No other genetic therapists have jumped the gun so far, but the ethics of genetic therapy started off on the wrong foot.

More serious than this mishap, however, has been an apparent decision by leading biologists not to participate in the public debate about human genetic engineering, presumably for fear of stirring up an imbroglio similar to that which surrounded the issue of hazard. This tactical decision makes a great deal of sense if, having learned nothing from the past, one believes that an ignorant public is safer than one in a position to give its informed consent. Its direct consequence is a climate of opinion in which the entire spectrum of church leadership can sign the statement discussed above that calls for a ban on making any inheritable changes to the human genome.

The Future Outlook

What is the outlook for the future? It is obvious that the more powerful biology becomes, the more its uses and the control of those uses will be debated. If biologists are perceived to be using their science for their own ends or abusing public trust, they will inevitably lose part or all of their autonomy. How much better for everyone if biology were to remain a community of autonomous, independent researchers, untrammelled by excessive commercial ties, free to give objective advice to whoever wants it, and interested only in the disinterested pursuit of truth. That's not the way things are headed at present, but the admirable sense with which biologists have conducted their affairs hitherto will doubtless prevail in the end.

18

Human Genetic Engineering

Alexander Morgan Capron

We are now in the middle period of public policy on genetic manipulation. Like the first stage, in which attention focused on the safety of laboratory experiments that altered the genes of microorganisms, the concerns at present involve safety, though on a larger scale: the creation of large quantities of recombinant DNA through industrial fermentation techniques for agricultural, manufacturing and pharmacological purposes, or the intentional release into the environment of plants and animals (beginning with bacteria) whose genes have been modified.[1] These developments have also spurred attention to social policy issues that underlie many areas of scientific investigation, such as the propriety of academic researchers and institutions having various connections with business enterprises, and the need to balance rewards to research sponsors (through patents and trade secrets) with fair access to the fruits of research on the part of the public, especially when public funds have contributed to the discoveries and when the benefits they provide may be matters of life and health for the people who need them.[2]

The unresolved status of many of the issues confronted during the first two stages will not prevent the arrival of the third stage, however. That stage — the extension of gene-splicing techniques to human beings — will be brought about not because public policy and ethical theory are ready for it, but because advances in biomedi-

This article is adapted from the David B. Brin Lecture, presented December 15, 1983, at the Johns Hopkins School of Medicine.

cal knowledge make it possible. Yet, unlike biomedical developments a decade or
two ago, such as heart transplants, which burst upon the public unawares, the coming manipulation of human genes has been discussed for many years, among not
only physicians and scientists, but also theologians, social scientists, philosophers,
lawyers and members of the general public in meetings and in the popular media.[3]

Despite years of prior discussion, the barriers to resolving the questions in the
third stage are even more formidable than those confronting participants in the
first two stages. Divisions of opinion in the recombinant DNA debate thus far have
turned on factors that are common in the formulation of public policy, namely,
"occurrence uncertainty": what is the probability of certain things occurring and,
given the likely consequences, who will benefit from, or be hurt by, different policies? Human genetic engineering also involves questions of occurrence uncertainty,
but it presents two sets of questions not encountered in the early stages: ethical uncertainty and conceptual uncertainty. As the President's Commission observed, the
first occurs "when no societal consensus exists as to whether certain applications of
gene splicing are beneficial or undesirable."[4] Beyond the question whether a particular human use — like improving memory — would be socially and ethically desirable, even more basic questions are encompassed within ethical uncertainty because
the determination of what constitutes a "defect" is itself not a definite concept,
fixed for all times or all people.

The second new type of uncertainty "refers to the fundamental change in concepts that this new technology can engender."[5] For example, the opening of realms
of therapeutic possibilities by the new genetics also destroys much of the security
that had been provided by fixed landmarks; the exchange and manipulation of
DNA among species challenges basic concepts and assumptions about what it
means to be human.

Complex Issues

These added layers of uncertainty are among the important ways that human genetic engineering raises more complex ethical and public policy issues than even
the highly contested issues that have already emerged in the pharmacological, industrial and agricultural research and development in recombinant DNA.

Of course, for the near future, issues of the latter type will continue to dominate
discussions, because most of the direct human uses will be some years in coming.
Nonetheless, those uses are not mere hypotheticals, and the issues they raise should
not be dismissed simply because they are not yet as pressing or apparently prevalent
as the nonhuman applications. Concern with those issues may partially result from
the technical wizardry involved, but it can also be traced to worries about the potential misuses of these techniques. Furthermore, some people doubt the morality
of any genetic manipulation in human beings or, at the least, believe that the burden of answering many concerns rests with the proponents of such efforts, which
ought not to go ahead in the meantime.[6] These ethical and social concerns — about
unacceptable uses or consequences, and about human gene splicing *per se* — will be
addressed after a brief description of the medical background.

Medical Uses of Gene Splicing

Remarkable progress has been made in improving health in the western world during this century. The initial triumphs resulted from public health measures, such as improved sanitation, water supply and immunization programs (as well as better food processing, nutrition and housing, which may not usually be considered matters of "health," although they have an enormous influence on it). More recently, dramatic steps have been taken to reduce the toll of trauma and acute diseases. The consequence is both a great lengthening of the average lifespan and improved functioning and enjoyment of life for many people. A further consequence, however, has been to throw the spotlight on the congenital, degenerative and chronic conditions, which now move to center stage as the major causes of disability and death in the United States. Since many of these conditions are "genetic"—or, at least, have a large genetic element—the search for diagnostic methods and treatments aimed at the genetic level will have increasing social as well as individual importance in the years ahead.

The techniques of genetic engineering—or "gene splicing"—are first being applied to human genetic diseases through a traditional route, namely, the production of enzymes and proteins for exogenous treatment of genetically based diseases like diabetes or growth failures by means of recombinant DNA-produced insulin and human growth hormone, respectively. Once the relevant gene can be identified, extracted from its place in the nucleus of human cells, spliced into a microorganism, and then caused to manufacture its gene product, a new way to produce plentiful quantities of formerly hard-to-obtain material is a reality. The ability of bacteria (often the "host" for a recombinant DNA chain) to multiply rapidly—a single cultured cell can become a billion copies in less than 15 hours—would seem to endow the gene-splicers with a biological Midas touch.

Gene Splicing Techniques

As formidable as this capability may seem, however, it is still rather conventional compared with more direct use of gene splicing techniques in diagnosis and manipulation of human genetic diseases. In the diagnostic area, the technique holds great promise for genetic disorders or carrier status that until now have not been readily diagnosable because present testing methods look for gene products, rather than for the genes themselves, and such biological markers are not always discernible.

The techniques can be briefly sketched.[7] Basically, they involve the use of enzymes as scissors to cut DNA chains and then the use of various methods (such as gel electrophoresis) to discern differences in the resulting pieces of DNA. In some cases, as has already been shown with some hemoglobin disorders, a restriction enzyme site occurs right at the relevant gene, while in the variant ("defective") gene the restriction site is absent, so that when the DNA is cut, a segment of a different length is produced, which indicates the presence of the defective gene. In other cases, the cutting-point is in a stretch of DNA adjacent to the gene of interest; then it is necessary to do familial studies to find this linkage. Recently, for example, a research team headed by James Gusella of Harvard announced an absolutely stunning use of the molecular genetic techniques — specifically, the use of restriction fragment length polymorphisms — to detect a marker on Chromosome 4 that appears to be closely linked to the gene for Huntington's disease.[8] This is a rare progressive neurodegenerative disorder with autosomal dominant inheritance; the first symptoms usually occur after the patient's child-bearing years. The primary defect is not yet known, but the symptoms — progressive motor abnormality and intellectual deterioration — are well studied, and are clearly devastating, leading to an early death. Although the potential to do presymptomatic — indeed, even prenatal — screening will raise ethical problems (about confidentiality, and particularly about autonomy), it is also plainly a great step forward in identifying the Huntington's gene itself and understanding its mechanism.

Besides diagnosing genetic defects, gene splicing may also make it possible to cure them — not in the way insulin injections "cure" diabetes, by counteracting its harmful effects on the patient, but by providing the genetic blueprint for a previously diabetic patient's body to make its own insulin. If this means of treatment — sometimes termed gene therapy or gene surgery — proves possible in human beings, it would offer a wholly new and potentially much less expensive way of treating many illnesses; if it makes inheritable changes, it would move another step beyond any current treatment (which is addressed to individual patients) and would become the treatment of future generations as well.

Such steps have not yet been successfully undertaken, but they are coming closer every day. Several recent developments are particularly important regarding human applications. First, because scientists do not yet fully understand how genes are regulated, it has been difficult to induce expression of foreign genes inserted into human cells — that is, to get the gene to function and program the cell to produce the relevant protein. Remarkable progress has been made lately, however, and scientists have been successful in getting recombinant genes to function in multicell animals and even in correcting a gene defect — a feat accomplished in fruit flies through the discovery of what are termed "transposable elements."[9] Although the counterparts of these elements have not yet been found in man, this experiment points the way to the future therapeutic uses of gene splicing. Second, there is the question: Can the changes brought on by gene splicing be passed on to subsequent generations? In experiments with mice, this important effect has now been demonstrated.[10]

The Two Categories

Gene splicing in humans can be divided into two categories: *somatic cell treatment* and *germ cell treatment*. It is likely that somatic cell treatment will initially be directed at single gene mutations, which are now known to cause more than 2,000 human disorders. For disorders that involve an identifiable cell product of a discrete subpopulation of cells, treatment might consist in removing some or all of these cells, genetically altering their genes, and then reinserting the cells into the patient. Thus, the technique is in effect organ transplantation — except that the physicians would alter a patient's own organ, rather than using another human being's organ.

From the viewpoint of medical ethics, this may be no small matter, because of the greater uncertainties, but conceptually it seems to me no different—provided that we have agreement on what constitutes a "disease" for which "therapy" is appropriate. Indeed, the criticism that was directed at the first attempt to use recombinant DNA techniques to treat patients with a genetic disease was of the same type that could be leveled at any experiment undertaken prematurely and without the necessary approval of an institution's "human subjects" review board.

That experiment, conducted by Dr. Martin Cline of the UCLA Medical School on two thalassemic patients in Israel and Italy in 1980, was condemned, and Dr. Cline was punished by the National Institutes of Health's cancellation of research funding, because his laboratory results and animal tests did not demonstrate sufficient likelihood of success to justify going ahead clinically.[11]

Since then, other researchers have pressed ahead, both to develop better animal "models" of the hemoglobin disorder, and to find ways around the problems that kept the altered DNA in Dr. Cline's experiment from having a beneficial effect. It seems likely that more experiments will take place in the near future, and that the initial experiments are likely again to involve an organ (like bone marrow) that can be physically removed, altered genetically, and returned to the patient's body.[12]

If only certain cells were altered, the recombinant DNA would be limited in its effect to those cells, and the individual's genetic material would otherwise remain the same as it always had been. Were an attempt made to change a person's genes to involve other methods of introducing the new genetic information, however, a large number of cells, including the germ-line cells, could be affected. This might occur if the new genetic material were introduced systemically (because the affected organ cannot be removed for separate treatment outside the body) or if the disease in question had to be treated early in development, because it affects many organs or is manifested irreversibly before birth.

For example, if the gene splicing took place at the zygotic stage, it would probably affect all cells in the body. Although treatments of this sort are almost certainly further in the future than other therapeutic uses of gene splicing, they raise much more troubling issues. First, there are the special ethical problems of creating human beings with the intention of altering them—entirely without their consent.[13] Second, social and biological concerns arise because any deleterious changes, rather than being limited to one person or one generation, would become part of the human genetic inheritance.

As important as social and ethical issues of this sort are, their mere recitation ought not to be substituted for careful analysis; valid objections to particular uses of gene splicing need not provoke wholesale rejection of the techniques themselves. It is worthwhile remembering that, a little more than a decade ago, the field of genetics was seen by many of its practitioners to have reached a plateau; problems remained to be explored, but mostly at the interstices, not the frontiers.[14] Gene splicing changed all that, and the exploration of gene functioning in higher organisms that splicing permits "turns out to be full of surprises."[15] The knowledge produced—about everything from the theory of evolution to the process of aging—may hold more for science than the practical applications of gene splicing hold for technology, at least in the near future. Therefore, calls for bans on areas of research

—which could easily chill other types of research — ought to be greeted very skeptically.

Decision-makers should remember that the key that they are told to put out of reach may be the one that could open storehouses of knowledge of unimaginable dimensions. Thus, the mere possibility that the key may also fit a biological Pandora's box — while enough to demand that public and scholarly attention be paid — is not enough to impose a moratorium on carrying the work forward into the medical realm.

Concerns with Consequences

A wide range of potentially troublesome results have been raised by critics of human genetic engineering.[16] Four deserving special attention are the effects on (a) human genetic makeup; (b) intergenerational responsibilities; (c) distribution of social benefits, and (d) the conception of what a person is.

Evolution and the Human Genome

In the early years of recombinant DNA research on microorganisms, many critics shared the view expressed by biologist Erwin Chargaff, who challenged the right of workers in this field "to counteract, irreversibly, the evolutionary wisdom of millions of years."[17] The objection focused on the notion that the inability of different species to mate and produce fertile offspring must offer a natural protection that scientists should not circumvent, lest they destroy an adaptive advantage.

There are several problems with this view, however. First, the scientific theory of evolution contains no notion of a plan or endpoint, so "wisdom" is a misleading metaphor — one set of genes is not necessarily more desirable than another. Even genes well adapted to a particular time and place may be maladaptive — perhaps lethal — under other circumstances. Whatever role "species barriers" played in evolution thus far would not mean that they are necessarily of continuing value.

Furthermore, the ability to move genetic material readily might itself become the means of ensuring species survival if, as philosopher Stephen Stich has argued, a time could come "when, because of natural or man-induced climatic change, the capacity to alter quickly . . . genetic composition" will be needed to forestall a catastrophe.[18]

This is not to say that all concerns about effects on the human gene pool should be dismissed. While it is not possible to know which particular genes would prove advantageous in a changed environment, population geneticists regard the loss of even minute advantages as serious, since the cumulative effect over generations can give a species marked benefits. Moreover, in the face of a sudden change in the environment (such as the introduction of a novel pathogen), a species is more likely to survive if its members possess greater heterogeneity. It seems unlikely, however, that therapeutic interventions aimed at eliminating genes in a form that is deleterious to an individual (*e.g.,* a person who is homozygous for an autosomal recessive disorder) would have a significant impact on the rate of the gene's occurrence in the population, especially given how rare most deleterious genes are. (This would also be true if genetic alterations of germ-line cells were limited to affected individuals; obviously, a much more dramatic impact on a gene's frequency would follow in the unlikely event that the germ cells of *all* carriers were altered—an issue that is also raised by screening aimed at carriers of dominant disorders.)

Medical geneticists tend to be much less concerned with any of these changes than are population geneticists, "because they believe that it should be possible to make up, through environmental manipulation (including medical treatment), for the loss of any advantage provided by a variant in any probable future environment."[19] Nonetheless, as scientists (aided by the tools of molecular genetics) learn more about the beneficial effects of gene variants—distinct from other environmental and inherited causes—it would seem desirable to take the diminution of the frequency of such variants into account in weighing the costs of a gene splicing program.

Intergenerational Responsibilities

Human genetic manipulation would add considerably to the challenges that other developments in reproductive and genetic medicine have already presented to traditional notions of parental and societal obligations toward children. In some circumstances, these obligations may seem to be expanded; in other ways, contracted.

Already the ability to predict genetic defects in unborn children has led some prospective parents to choose to terminate a pregnancy (and perhaps to "try again"). For the family involved, this result may be a great blessing, justified in ethical as well as practical terms, especially when the fetus that was aborted would have had a painful existence. Nonetheless, the rapid acceptance of prenatal diagnosis ought not to obscure the fact that, when used with selective abortion, it upsets the traditional norm that children are to be accepted unconditionally, even when their birth creates a burden, like one of life's other mysterious tragedies. To the extent that gene splicing greatly expands the range of prenatal diagnoses, it will accelerate the rejection of traditional attitudes and reenforce a growing sense that human imperfections need not be tolerated.

The use of genetic engineering for therapeutic rather than merely diagnostic ends could have even more far-reaching effects on people's links to, and responsibilities for, their progeny. It may no longer be seen as appropriate for "responsible parents" simply to accept the result of the natural lottery by which characteristics are now determined; instead, they may be expected to "correct" genes that cause diseases and to "augment" other genes to give their children opportunities for higher levels of physical or cognitive functioning. On the other hand, knowing that future generations may employ an even more advanced technology to alter or to replace characteristics passed on to them could weaken people's sense of genetic continuity.

Furthermore, by blurring the line between what counts as a serious defect or disability and what is "normal functioning," gene splicing may alter our perception of what society owes to children, particularly those burdened by handicaps. Today's norm may become tomorrow's deficit; those problems that could have been genetically corrected at some prenatal or even preconceptual point may be seen as matters of human choice (comparable to the problems or disadvantages that result from a wide range of parental choices about their children's schooling, health care, and general upbringing) and hence less demanding of the beneficial or charitable impulses of society.

The interrelationship between genetics and social and psychological behavior and attitudes is complex and poorly understood. Yet it is apparent that people's impressions of sharing constitutional similarities with their kin reenforces family solidarity and a sense of mutual obligation. "If genetic engineering makes use of reproductive technologies such as artificial insemination and *in vitro* fertilization, it will increase the strains on this concept of lineage."[20] Consequently, the possibility of using genetic technologies to correct defects creates uncertainties both about what effects on intergenerational relationships are likely and about how to evaluate their ethical effects.

Distribution of Social Benefits

In the third area of possible effects, one moves from those that illustrate ethical uncertainty to those that also raise questions of conceptual uncertainty. The core ethical concern is that of fairness. Human gene splicing confounds our attempts at distributive justice at two levels: What is distributed, and who deserves it.

Most of medicine — and all that seems heroic or praiseworthy — aims to hold back death, relieve suffering, and overcome disability. The things that provoke the need for medical intervention are viewed as disruptions in our personal universe and labeled as such — scourges, plagues and disasters (or, in the cooler terms of medicine, diseases, disorders and syndromes) — whether they originate within the body or as the result of some external contagion or accident. Thus, the targets of medicine are relatively clear to the practitioner, researcher or bureaucrat, as well as to the ordinary citizen-patient.

The potential of genetic engineering (admittedly, a more remote potential than its application against single-gene defects) to alter "adequate" or "normal" functioning would cast medicine into an uncomfortable role. To the extent that it has already been cast into that role on occasion, it seems obvious that those occasions — for example, cosmetic surgery or prescription of drugs to enhance athletic performance — have been marked with controversy and a general sense of dis-ease by many in the field. And even those earlier examples have been grounded in a firm sense of what was normal and a recognition of the artificiality of manipulation.

Gene therapy blurs such lines. While it may be "normal" (in the sense of inheritance of certain genes from a person's parents) for a person to be afflicted with sickle-cell anemia, it is not "normal" for the population as a whole, and most people with this genetic makeup will not experience an adequate level of health without medical intervention, if even then. The replacement of the functioning of the sickle-cell genes with genes that would program the production of normal hemoglobin would seem an obviously beneficial use of gene splicing.

But it is also not normal, in the sense of *average,* for a person to have 160 IQ or to be able to run 100 meters in under 10 seconds. It would, however, be surprising were someone to suggest that, when "abnormal" genes occur that provide the bases for these abilities, they should be removed. But what about the reverse — for example, giving the hypothetical genes for mental acuity to someone who has an IQ of 100, to push the person to 160? If a society is deciding where to expend funds for research and treatment, should such an alteration be included among those things that a society that desires to be fair will encourage — or prohibit? Or neither? How, indeed, can one conceptualize any notion of equity in the area of health if the notion of some adequate level (in terms of both resources and outcome) is wholly indeterminate?

Even if it were agreed that an increase in intelligence or athletic abilities were a medical good, like a means of overcoming sickle-cell anemia, that ought to be made available, who would deserve to get it? This is the second problem posed for anyone seeking to be fair about human gene splicing. The usual answer to such a problem is to seek a means of distribution that is either rationally related to the good being distributed, or that is completely neutral (such as a lottery). The latter would seem an odd way to distribute a form of medical treatment (or "enhancement"), yet the former is likely to be impossible, as Michael Shapiro has noted:

> What if intelligence could be engineered upward? Who would merit this increase in merit? The very oddity of the inquiry calls into question the continued use of intelligence as a basis for resolving competing claims — say, for admission to educational institutions or for access to the intelligence-raising technology itself.[21]

Thus, if genetic engineering ever moves beyond the treatment of generally recognized diseases and comes into use to alter more complex characteristics, it will pose the very old problem of distributive justice in a very novel, perhaps unanswerable way. Indeed, it could "call into question the scope and limits of a central element of democratic political theory and practice: the commitment to equality of opportunity."[22] Would true commitment to equality require in the end mandatory genetic alterations to give everyone the same (or equally functional) genes?

The Concept of Being Human

The notion of such radical changes in people's genetic makeup raises yet another set of social and ethical issues, namely the challenges that genetic engineering poses for the concept of being human. This has several facets. On the individual level, a person's sense of identity could be called into question by genetic changes; more generally, changes of degree at some point become changes in kind.

The first of these possible effects might be viewed simply as a matter of change in psychology. Although people now think of themselves (and others) as relatively fixed in their capabilities, characteristics and personalities once they pass adolescence, dramatic changes are not unknown (as unwelcome as they often are to those who have anticipated a certain stability in their relationships with others). But even such changes are either viewed as a further manifestation of a person's true inner self or, when they are not, are resisted by others and perhaps subjected to medical intervention, designed to restore the person to his or her "true self." Compared to the methods now available, genetic engineering could well be faster and more selective, and almost certainly could extend well beyond anything now attempted through psychotherapy (including psychopharmacology).

> Here again, uncertainty about possible shifts in some of people's most basic concepts brings with it evaluative and ethical uncertainty because the concepts in question are intimately tied to values and ethical assumptions. It is not likely that anything so profound as a change in the notion of personal identity or of normal stages of development over a lifetime is something to which people would have clear value responses in advance.[23]

The second and more general change in the concept of being human comes down to the ancient question: What is human nature? Can it be described by those characteristics that are uniquely human? This would appear to be a very narrow category, since most "human genes" and many "human characteristics" are found in other species. Would the addition of new capabilities to an otherwise human creature render it nonhuman? What of a genetic subtraction of a few capabilities — such as the ability to record and study the past and plan beyond the immediate future?

The very notion of a *direction* of change implicit in the preceding questions itself poses a problematical point: that humankind knows what an improvement or a degradation in its nature would be. The notion of betterment has long been attractive to geneticists. Herman J. Muller was one of the leading scientific spokesmen for this view; his notion of the desirable genetic traits changed with fashions in politics, however.[24] The coming of a means to manipulate the genes directly (rather than through the random chances that attached to Muller's proposals for selective breeding) offered human beings the opportunity to "rise above their nature," as Robert Sinsheimer observed in the late 1960s.[25]

Nonetheless, in time, Dr. Sinsheimer came to doubt the wisdom of many proposed uses and investigations of genetic engineering.[26] Even if the power to make genetic changes were not seized by an evil government (always a danger with any great power — but the fault is in the powerful government, not in the science it seizes), it might cause harm despite its user's benign intent. Indeed, 20 years ago, C.S. Lewis described the arrival of "one dominant age" that could overcome all influences of the past while simultaneously pre-ordaining the actions and capabilities of all subsequent generations.

> Man's conquest of Nature, if the dreams of the scientific planners are realized, means the rule of a few hundreds of men over billions upon billions of men.[27]

It is by no means certain that it will ever be possible to change the genetic basis of all or even the most important human characteristics in a predictable, inheritable way. Compared to the immediate threats of nuclear holocaust and ecological degradation, the social and ethical difficulties posed by developments in human genetic engineering seem remote and perhaps insignificant. Yet events in science have a way of overtaking the unwary. Although the large degree of uncertainty in outcome that marks each of the possibilities discussed here prevents their definitive evaluation in ethical and social terms, it is none too soon to begin attending to the important matters of ethical and conceptual uncertainty raised by the mere possibility that physicians will soon be able to make direct — and directed — genetic changes in human beings.

The Need for Circumspection

Science fiction is seldom great literature, and since Nathaniel Hawthorne is, without question, a great writer, it is not surprising to discover that, while one of his stories may look, to our modern eyes, like science fiction, it is actually a moral tale. Nonetheless, the morality carries a large message about scientific ethics, especially as applied to the potentialities of human gene splicing.

The story I have in mind concerns events in the life of a young man, Giovanni Guasconti, who went many years ago from the south of Italy to study at the University of Padua. The garret he rents overlooks the garden of Dr. Rappaccini, an eminent physician, whom Giovanni observes tending his strange flowers and plants with the assistance of his lovely daughter, Beatrice. Pietro Baglioni, another eminent professor of medicine at the university, advises Giovanni to avoid Rappaccini, who "cares infinitely more for science than for mankind." In words that ring with greater resonance today than they can have in Hawthorne's time, Professor Baglioni says that Rappaccini's

> patients are interesting to him only as subjects for some new experiment. He would sacrifice human life, his own among the rest, or whatever else was dearest to him, for the sake of adding so much as a grain of mustard seed to the great heap of his accumulated knowledge.

Giovanni—partially, one senses, out of the irresistible attraction he already feels for Rappaccini's daughter—objects that this reveals a "noble spirit." "Are there many men," he wonders, "capable of so spiritual a love of science?"

Yet Hawthorne seems to take a harsher view of the scientist, for in the end Dr. Rappaccini's research leads to tragedy. His work—in what seems so prescient for today—involves the creation of artificial forms of life, albeit not (so far as we are told) through genetic engineering. The plants in Rappaccini's garden display

> an appearance of artificialness indicating that there had been such commixture, and, as it were, adultery, of various vegetable species, that the production was no longer of God's making, but the monstrous offspring of man's depraved fancy, glowing with only an evil mockery of beauty.

Like the plants in her father's garden, lovely Beatrice turns out to be possessed of man-made qualities, and, like them, it turns out that the central quality is being poisonous—her sweet breath literally kills insects, and ordinary flowers wilt in her hands.

The power possessed by Beatrice is so great that her father has kept her closed off from the world in his garden (which one might thus liken to the first P-4 laboratory!). Wishing to overcome Beatrice's total isolation from people, Dr. Rappaccini alters Giovanni to be like her; when Giovanni discovers this, he turns on Beatrice and condemns her—an early example of what is now called "blaming the victim." Finally, Beatrice and Giovanni—wishing to become normal humans again—decide to take an antidote supplied by Professor Baglioni.

Perhaps sensing the danger for a person with poison taking an antidote, Beatrice insists on going first. She rebuffs her father's claim that she has been "endowed with marvelous gifts against which no power or strength could avail an enemy," and then she dies, "the poor victim of man's ingenuity and of thwarted nature, and of the fatality that attends all such efforts at perverted wisdom," as Hawthorne writes.

It is hard to know whom to pity more — poor Beatrice, who is dead; Dr. Rappaccini, who has lost his precious child, and perhaps his experiment as well; Giovanni Guasconti, who is left facing life alone in the poisonous garden — or death at his own hands; or perhaps Professor Pietro Baglioni — in the role of societal watchdog — who was able to prevent generalized harm from arising from the experiment only by a step that led to the death of its subject, the lovely Beatrice.

In a way, Hawthorne's "Rappaccini's Daughter" has more to say of relevance to the situation of gene splicing than does Mary Shelley's story of Dr. Frankenstein and his monster, which was the dominant metaphor of the early years of gene splicing. We may have less to fear from a monster run amok than from the unfavorable effects of our successes. But we need not suffer Dr. Rappaccini's fate, if we remember the human values and goals that ought to guide the human uses of genetic engineering — just as they do elsewhere in that application of science and art to human needs which we call medicine.

Notes

1. *See generally* M. Rogers, *Biohazard* (1973); N. Wade, *The Ultimate Experiment* (1977); "NIH Guidelines for Research Involving Recombinant DNA Molecules," *Fed. Reg.* 41 (July 7, 1976), p. 27902 (first version of frequently revised rules on experimental safety); Fox, "Despite Doubts RAC Moving to Widen Role," *Science* 223 (1984), p. 798.
2. *See e.g.,* the article by Peter Hutt in this series for *Technology In Society* (Vol. V, no. 2, pp. 107–118); *Commercialization of Academic Biomedical Research,* Hearings Before the Subcommittee on Investigation and Oversight and the Subcommittee on Science, Research and Technology of the House Committee on Science and Technology, 97th Congress, 1st Session, June 8, 1981; Culliton, "Pajaro Dunes: The Search for Consensus," *Science* 216 (1982), p. 155.
3. *See generally* M. Hamilton, ed., *The New Genetics and the Future of Man* (1972), pp. 109–175; M. Lappé and R. Morison, eds., *Ethical and Scientific Issues Posed by Human Uses of Molecular Genetics, Annual of the New York Academy of Science* 265 (1976) pp. 1–208; J. Little, ed., *Prospects for Man: Genetic Engineering* (1979); Roblin, "Human Genetic Therapy" in G. Chacko, ed., *Health Handbook* (1979), p. 103; Anderson and Fletcher, "Gene Therapy in Human Beings: When Is It Ethical to Begin?," *New England Journal of Medicine* 303 (1980), p. 1293; J. Cherfas, *Man-Made Life* (1982), pp. 228–234.
4. President's Commission for the Study of Ethical Problems in Medicine and Biomedical and Behavioral Research, *Splicing Life* (Washington, DC: U.S. Government Printing Office, 1982), p. 22.
5. *Ibid.*
6. This appears to have been the position of the three religious organizations whose objections to President Carter about the "fundamental danger triggered by the rapid growth of genetic engineering" led to the request that the President's Commission study the field. See *President's Commission, supra,* note 4, pp. 1, 95–96 (Appendix B). *See also Human Genetic Engineering,* Hearings Before the Subcommittee on Investigation and Oversight of the House Committee on Science and Technology, 97th Congress, 2nd Session, November 16–18, 1982 (especially comments of R. Shinn, R. McCormick and J. Beckwith).
7. *See generally ibid.*, pp. 238–278 (comments of Y. Kan and M. Skolnick); Wyman and White, "Restriction Fragment Length Polymorphism in Human DNA," *77th Proceedings of the National Academy of Sciences* (1981), p. 6754.
8. Gusella *et al.,* "A Polymorphic DNA Marker Genetically Linked to Huntington's Disease," *Nature* 306 (1983), p. 234.
9. Marx, "Still More About Gene Transfer," *Science* 218 (1982), p. 459.
10. Wagner *et al.,* "Microinjection of a Rabbit B-globin Gene Into Zygotes and Its Subsequent Expression in Adult Mice and Their Offspring," *78th Proceedings of the National Academy of Science* (1981), p. 6376.

11. Williamson, "Gene Therapy," *Nature* 298 (1982), pp. 416, 418; and Wade, "UCLA Gene Therapy Racked by Friendly Fire," *Science* 210 (1980), p. 509.

12. *See generally* Roblin, *supra,* note 3.

13. R. Ramsay, *Fabricated Man* (1970).

14. Cherfas, *supra,* note 3, pp. 24–25.

15. Judson, "Thumbprints in Our Clay," *The New Republic,* September 19 & 26, 1983, p. 12, 15:

> The problems for which the techniques of genetic engineering are indispensable are the most interesting in biology, perhaps in all of science. This technology is popularly presented as irresistible for its practical benefits. Far more significantly, it is irresistible for fundamental science.

16. *See generally* J. Rifkin, *Algeny* (1983).

17. Quoted in Cavalieri, "New Strains of Life—Or Death," *The New York Times Magazine,* August 22, 1976, pp. 8, 68.

18. Stich, "The Recombinant DNA Debate," *Philosophy and Public Affairs* 7 (1978), p. 187.

19. President's Commission, *supra,* note 4, p. 64.

20. *Ibid.,* p. 65.

21. Shapiro, "Introduction to the Issue: Some Dilemmas of Biotechnology Research," *Southern California Law Review* 51 (1978), pp. 987, 1001–02.

22. President's Commission, *supra,* note 4, p. 67.

23. *Ibid.,* p. 68.

24. Allen, "Science and Society in the Eugenic Thought of H.J. Muller," *BioScience* 20 (1970), p. 346.

25. Sinsheimer, "The Prospect of Designed Genetic Change," *Engineering and Science* 32 (April 1969), pp. 8, 13.

26. Dixon, "Tinkering with Genes," *Spectator* 235 (1975), p. 289.

27. C.S. Lewis, *The Abolition of Man* (1965), p. 71.

19

Oswald Avery and the Cascade of Surprises

Lewis Thomas

One of the liveliest problems in cancer research these days is the behavior of onco-genes, strings of nucleic acid that were originally found in several of the viruses responsible for cancer in laboratory animals, later discovered to exist in all normal cells. By itself, an oncogene appears to be a harmless bit of DNA, maybe even a useful one, but when it is moved from its normal spot to a new one on another chromosome, it switches the cell into the unrestrained growth of cancer. It is not yet known how it does this, but no one doubts that an answer is within reach. Some of the protein products of oncogenes have already been isolated and identified, and the site of their action within the cell, probably just beneath the cell membrane, is now being worked on. The work is moving so fast that most recent issues of *Science* and *Nature* contain several papers on oncogenes and their products.

But cancer is only one among dozens of new problems for the scientists who work on DNA. The intimate details of genetically determined diseases of childhood are coming under close scrutiny, and the precise nature of the error in the DNA mole-cule has been observed in several, sickle-cell disease, for example, with more to come. The immunologists, embryologists, cell biologists, and, recently, even the endocrinologists have observed the ways in which small modifications of DNA affect the behavior of the systems that they study. *Observed* is exactly the right word here: the DNA molecule can be inspected visually, purified absolutely, analyzed chemically, cut into whatever lengths you like, transferred from one cell to another or into a test tube for the manufacture of specific proteins. The techniques and in-struments now available for studying DNA have become so sophisticated that I recently heard a young colleague, one of the new breed of molecular geneticists, complain that "even a dumb researcher can do a perfectly beautiful experiment."

It is indeed a biological revolution, unquestionably the greatest upheaval in biology and medicine ever.

It began just fifty years ago in a small laboratory on the sixth floor of the Hospital of the Rockefeller Institute for Medical Research, overlooking the East River at Sixty-sixth Street in New York. Professor Oswald T. Avery, a small, vanishingly thin man with a constantly startled expression and a very large and agile brain, had been working on the pneumococcus — the bacterium causing lobar pneumonia — since the early years of World War I.

Avery and his colleagues had discovered that the virulence of pneumococci was determined by the polysaccharides (complex sugars) contained in the capsules of organisms and that different strains of pneumococci possessed different types of polysaccharide, which were readily distinguished from one another by specific anti-bodies. It was not clear at that time that the type of polysaccharide was a genetic property of the bacterial cells — it was not even clear that genes *existed* in bacteria — but it was known that the organisms bred true; all generations of progeny from a pneumococcus of one type were always of that same type. This was a solid rule, and like a good many rules in biology there was an exception.

An English bacteriologist, Fred Griffith, had discovered in 1923 that pneumo-cocci could be induced to lose their polysaccharides and then to switch types under special circumstances. When mice were injected with a mixture of live bacteria that had lost their capsules (and were therefore avirulent), together with heat-killed pneumococci of a different type, the animals died of the infection, and the bacteria recovered from their blood were now the same type as the heat-killed foreign or-ganisms.

Avery became interested in the phenomenon and went to work on it. By the early 1930s it had become the main preoccupation of his laboratory. Ten years later, with his colleagues Colin M. MacLeod and Maclyn McCarty, the work was completed, and in 1944 the now classic paper was published in the *Journal of Ex-perimental Medicine,* formidably entitled "Studies on the Chemical Nature of the Substance Inducing Transformation by a Desoxyribonucleic Acid Fraction Isolated from Pneumococcus Type III."

The work meant that the genes of pneumococci are made of DNA, and this came as a stunning surprise to everyone — not just the bacteriologists but all biolo-gists. Up until the announcement, the concept of the gene was a sort of abstrac-tion. It was known that genes existed, but nobody had the faintest idea what they were made of or how they worked. Here, at last, was chemical evidence for their identity and, more importantly, a working model for examining their functions.

Several years later a new working model was devised in other laboratories, involv-ing viruses as the source of DNA, and in 1953 the famous paper by James Watson and Francis Crick was published, delineating the double-helix structure of DNA. Many biologists track the biological revolution back to the Watson-Crick discovery, but Watson himself wrote, "Given the fact that DNA was known to occur in the chromosomes of all cells, Avery's experiments strongly suggest that future experi-ments would show that all genes were composed of DNA."

Looking back, whether back to the Watson-Crick paper or all the way back to Avery, the progress of science can be made to seem an orderly succession of logical steps, a discovery in one laboratory leading to a new hypothesis and a new experi-ment elsewhere, one thing leading neatly to another.

It was not really like that, not at all. Almost every important experiment that moved the field forward, from Avery's "transforming principle" to today's "jumping genes" and cancer biology, has come as a total surprise, most of all surprising to the investigators doing the work. Moreover, the occasions have been exceedingly rare when the scientists working on one line of research have been able to predict, with any accuracy, what was going to happen next. A few years ago certain enzymes were discovered that will cut DNA into neat sections, selectively and precisely, but it could not have been predicted then that this work would lead directly, within just a few more years, to the capacity to insert individual genes from one creature into the genetic apparatus of another—even though this is essentially what Avery accomplished, more crudely, to be sure, a half century ago.

Good basic science is impossible to predict. By its very nature, it must rely on surprise, and when it is going very well, as is the case for molecular genetics today, it is a cascade of surprises.

And there is another sort of surprise that is essential for good basic science, not so exhilarating, enough to drive many student investigators clean out of science. This is the surprise of being wrong, which is a workaday part of every scientist's life. Avery endured ten years of it, one experiment gone wrong after another, variables in the system that frequently made it impossible to move from one question to the next. Reading the accounts of those ten years in Rene Dubos's book *The Professor, The Institute, and DNA,* one wonders how Avery, MacLeod, and McCarty had the patience and stubbornness to keep at it.

Being wrong, guessing wrong, setting up an elegant experiment intended to ask one kind of question and getting back an answer to another unrelated, irrelevant, unasked question, can be frustrating and dispiriting, but it can, with luck, also be the way the work moves ahead. Avery was especially good at capitalizing on mistaken ideas and miscast experiments in his laboratory. He is quoted by some of his associates as having said, more than once, "Whenever you fall, pick something up."

This is the way good science is done: not by looking around for gleaming negotiable bits of truth and picking them up and pocketing them like game birds, nor by any gift of infallible hunch of where to look and what to find. Good science is done by being curious *in general,* by asking questions all around, by acknowledging the likelihood of being wrong and taking this in good humor for granted, by having a deep fondness for nature, and by being made nervous and jumpy by ignorance. Avery was like this, a familiar figure, fallible but beyond question a *good* man, the kind of man you would wish to have in the family. Accident-prone, error-prone, but right on the mark at the end, when it counted.

One thing for a good basic scientist to have on his mind, and worry about, is how his work will be viewed by his peers. If the work is very good, very new, and looks as if it's opening up brand-new territory, he can be quite sure that he will be criticized down to his socks. If he is onto something revolutionary, never thought of before, contradicting fixed notions within the community of science involved, he tends to keep his head down. When he writes, he writes as Avery wrote, a cautious paper, as noncommittal as possible, avoiding big extrapolations to other fields.

Not all scientists with great discoveries to their credit receive the Nobel Prize, and Avery did not. This spectacular omission continues to mystify the scientific community and has never been explained. Rene Dubos thinks that the Nobel committee was not convinced that Avery knew the significance of his own work, perhaps because of the low-key restraint with which the manuscript was written; the paper did not lay out claims for opening the gate into a new epoch in biology, although this is certainly what it did accomplish.

I doubt very much that Oswald Avery was ever troubled by the absence of a Nobel Prize or even thought much about it. He was not in any sense a disappointed man. He understood clearly, while the work was going on, what its implications were, and when it was finished he was deeply pleased and satisfied by what it meant for the future. He had set the stage for the new biology, and he knew that.

Contrary to the general view, not all scientists do their best work in their thirties, peak in their forties, and then subside. Avery was sixty-seven years old when his DNA paper was published. He retired four years later and died of cancer at the age of seventy-seven, at peace with the world.

Part VI

20
Introduction
Joseph G. Perpich

This final group of articles (which was not included in the *Technology in Society* series) identifies a major policy area of regulatory oversight that falls under the purview of the US Department of Commerce—namely, export controls. The Commerce Department's impending development of export regulations for biotechnology processes and products will have a direct impact on the biotechnology industry and, possibly, depending on their scope, on the academic community as well. Development of these export regulations is driven by the perceived need to monitor foreign industrial targeting of biotechnology and the possible misuse of biotechnology for purposes of biological warfare. Several government meetings on technology transfer have been held with biotechnology industry representatives during the past two years, including a White House symposium in April 1983 and a Commerce Department meeting in March 1984. Because of these policy developments, I organized and chaired a panel on biotechnology and international trade at the Annual Meeting of the American Association for the Advancement of Science (AAAS) in May 1984. That panel drew on the participants from these government meetings, and the ensuing articles represent the presentations of the panelists.

Clyde V. Prestowitz, Jr.

The first contributor is Clyde V. Prestowitz, Jr., Counselor to the Secretary of Commerce for Japan. Mr. Prestowitz is a graduate of Swarthmore College and holds a master's degree from the University of Hawaii's East-West Center and Keio University in Tokyo. In 1980, he earned an M.B.A. from the Wharton School of Finance. Mr. Prestowitz served for two years as a foreign service officer in Washington and the Netherlands. Following his government service, he began an international business career and subsequently formed Prestowitz Associates, a consulting group that assists companies in developing international strategies emphasizing Japan and technology transfer issues. Mr. Prestowitz joined the US Commerce Department in September 1981 as Deputy Assistant Secretary for International Economic Policy, and was appointed to his present position in August 1983.

In his paper, Mr. Prestowitz addresses the issue of foreign targeting in biotechnology. He begins by reviewing the growth of the new biotechnology firms which, with major corporations in virtually every industrial sector, have the potential for billions of dollars of product sales in the 1990s. These commercial possibilities have attracted the attention of government policymakers throughout the world who see this technology as a means for national industrial development. As others in this series have described, the US system for support of basic research at the universities has fueled this industrial revolution, but abroad, Mr. Prestowitz observes, governments foster development under far different conditions.

For example, biotechnology is almost exclusively a large corporate activity and has been targeted by the government for commercialization. Without anti-trust constraints, and through coordination by the Ministry of International Trade and Industry (MITI), Japan's corporations pool their resources and select which biotechnology sector each will pursue. Mr. Prestowitz cites US government concerns that these foreign targeting practices place US firms at a competitive disadvantage by limiting the necessarily free and open competition in the marketplace. Further, the capital and resources of the large foreign corporations are used to buy the technology from the newer biotechnology firms in the United States, resulting in major transfers of US technology. A key priority of the US government, Mr. Prestowitz concludes, is to challenge these practices, and, where possible, to seek their dismantling.

John H. Birkner

The second contributor is John H. Birkner. Dr. Birkner, a Colonel in the US Air Force, received his Ph.D. in environmental biology from Colorado State University, and served for several years on the faculty of the US Air Force Academy and as a foreign military technology analyst in the Defense Intelligence Agency. He is presently assigned to the Allied Forces South NATO staff in Italy.

In his presentation, Dr. Birkner outlines the efforts of the Defense Department in developing biotechnology guidelines for purposes of the Militarily Critical Technologies List (MCTL). This list is a compilation of technologies—products and processes—that Defense wants to monitor or to regulate. The list provides guidance to the Commerce Department in its implementation of the Export Administration Act of 1979, which empowers the Department of Commerce to regulate "dual-use" exports, those with both military and civilian applications.

Dr. Birkner states that the current focus of the biotechnology guidelines is on equipment related to physical containment for the purpose of biosafety and fermentation technology. The major impetus for the development of these guidelines is the Defense Department's concern that this technology may be transferred to the Eastern bloc and possibly misused for biological or chemical warfare purposes. The US government has repeatedly affirmed that the Biological Weapons Convention prohibits such research for biological warfare purposes. Dr. Birkner states that the Soviet Union reportedly has an active R&D program that is investigating and evaluating the utility of biological weapons in direct violation of the convention. Thus, the purpose of the DoD guidelines is to place restraints on biotechnology exports that could enhance the chemical or biological warfare offensive (or defensive) capabilities of potential adversaries. Dr. Birkner concludes by urging a full exchange of information between industry and government so that sensible regulations can be developed to meet both the national security concerns of government and the need of the biotechnology industry for commercial development and product exports.

Hank Mitman

The next contributor is Hank Mitman, Director of the Capital Goods and Production Materials Division of the US Office of Export Administration. A graduate of Gettysburg College who did graduate work in physics at Johns Hopkins University, Mr. Mitman served for several years in the US Army in aerospace research and development. He then joined the private sector where he was an engineering manager of international operations with a major US corporation for a number of years. He joined the Commerce Department as a technical data export licensing specialist in the Office of Export Administration and subsequently became Division Director.

Mr. Mitman describes the statutory mandates and regulatory responsibilities of the Office of Export Administration. The Export Administration Act of 1979 empowers the Department of Commerce to regulate "dual-use" exports, those with both military and civilian applications. Under the statute, the Commerce Department has developed export regulations that organize commodities for export into ten categories. Biotechnology currently falls into two of those categories related to products of certain fermentation processes and shipment of microorganisms. Mr. Mitman summarizes 1973–1974 Office of Science and Technology Policy (OSTP) and Commerce Department meetings with biotechnology representatives, and describes Commerce Department initiatives for developing biotechnology export regulations. Specific biotechnology areas for possible export control include products of biotechnology processes, technology transfer, biosensors, and biochips.

Mr. Mitman provides the rationale for export controls on biotechnology as based on "dual-use," military criticality, foreign availability, and effectiveness of transfer. In the development of these regulations, two major policy areas to be addressed are the nature of the controls on the products and processes of biotechnology and the boundaries surrounding basic research (exempted from the act), and technology transfer that falls within the scope of the regulations. Mr. Mitman concludes by noting that the goal of the Commerce Department is to balance the national concerns of technological transfer with the need to foster the development of the US biotechnology industry. To accomplish this goal, a close working relationship with the biotechnology industry is essential.

Irving S. Johnson

Irving Johnson is Vice President of the Lilly Research Laboratories, a division of Eli Lilly and Company. He earned his undergraduate degree and his Ph.D. in experimental biology at the University of Kansas. Dr. Johnson has spent much of his professional life at Eli Lilly and became Vice President in 1973. He played a leading role as an industrial participant in the development and subsequent revisions of the NIH guidelines for recombinant DNA research, and he is one of the key leaders in the industrial development of biotechnology.

In his article, Dr. Johnson agrees with Mr. Prestowitz that biotechnology has become an arena for economic competition among the industrialized nations of the world, largely as a result of strategies derived from national science policies. He cites the governmental and industrial research strategies of Britain and Japan, which include direct governmental financial support and favorable tax, patent, and anti-trust legislative initiatives. Dr. Johnson points out that the pioneering developments in recombinant DNA occurred in the United States. To maintain this competitive edge, however, he recommends that a number of steps be taken, including federal support for basic research, bioprocessing, and applied microbiology; clarification of health, safety and environmental regulation by the federal government; and con-tinued scientific oversight by the NIH's Recombinant DNA Program Advisory Committee.

Regarding export regulations, Dr. Johnson questions the value of restrictions, generally, on biotechnology equipment. He notes that the equipment is available from sources other than the United States, and he believes that export controls would only harm US industrial sales abroad. Dr. Johnson, however, recognizes the concerns expressed by Dr. Birkner on the misuse of biotechnology for purposes of biological warfare. There is a legitimate need for international oversight, but Dr. Johnson cautions that there are many pathogenic organisms occurring naturally that can wreak havoc without the need to resort to genetic engineering technology.

Nanette Newell

The final contributor is Nanette Newell, former Director of Research Administration at Calgene, Inc., a new biotechnology firm specializing in agricultural applications. Dr. Newell was awarded her B.S. in chemistry from Lewis and Clark College, and her Ph.D. in biochemistry and cellular and molecular biology from the Johns Hop-kins University School of Medicine. Dr. Newell served as a post-doctoral Fellow at the University of Wisconsin and subsequently as a Congressional Science Fellow. In 1982 she became the Project Director for the Congressional Office of Technology Assessment (OTA) study on biotechnology. On the basis of this study, the OTA is-sued a report in January 1984, *Commercial Biotechnology: An International Anal-ysis.* The report provides an excellent analysis of factors affecting the ability of the United States to compete worldwide in the developing biotechnology industry. Cur-rently Dr. Newell is an industrial biotechnology consultant in the San Francisco area.

In her paper, Dr. Newell focuses on the means for international technology trans-fer through access to world markets and international competitiveness. She uses the pharmaceutical industry as a representative sector of the biotechnology industry in which industrial applications are rapidly reaching the marketplace. She describes access to world markets through such mechanisms as licensing, joint ventures, manufacturing through a subsidiary, and direct export of products. Licensing is a preferred means of market access for the smaller companies who lack marketing and manufacturing expertise. Joint ventures become the preferred method for firms with marketing expertise, because less technology is transferred abroad. Wholly owned subsidiaries are extensively used in the pharmaceutical industry as an excel-lent access to foreign markets. Export of products transfers the least technology, but one then must face the "thicket" of tariff and nontariff barriers.

Although the United States has a leadership role in biotechnology, Dr. Newell notes that basic research discoveries and industrial applications are by no means limited to the United States. Thus, legislation and regulation governing export controls on technology must recognize the international scientific enterprise in biotechnology. Dr. Newell reports on the legislative stalemate in the reauthorization of the Export Administration Act. (As Mr. Mitman notes in his paper, the export regulations continued in effect during 1984 under the International Emergencies Economics Powers Act. Congress reauthorized the Export Administration Act in 1985.) Dr. Newell concludes that the government has increasingly focused on regulatory control of technology rather than on products. She cautions that this approach can impede scientific exchange and hinder US access to world markets.

In sum, the authors in this last series of articles highlight the national security and commercial policy considerations that will guide the Department of Commerce in its development of biotechnology export regulations. The policy debate over legislative and regulatory export control initiatives will be intense over the next several years, and the biotechnology industry will increasingly become a participant in that debate. As the authors in the series acknowledge, the biotechnology industry — and the academic community — must work with government to ensure biotechnology export rules based on legitimate commercial and national security interests that do not harm international exchanges upon which this industry and academic community so vitally depend.

21
Foreign Targeting in Biotechnology
Clyde V. Prestowitz, Jr.

Biotechnology represents a new frontier. Taken in its broadest sense, it has a long history, in fields ranging from fermentation to medicine. But within the last decade, it has assumed a new form and a new significance. Only within that period has the term itself even entered our vocabulary. But its widespread use today indicates its expected significance for science, industry, and agriculture.

A precise, comprehensive definition of the word is difficult, but in essence, biotechnology entails the application of genetic, cell fusion (hybridoma) and fermentation technology for industrial production. The key breakthroughs for industrial applications, in such areas as recombinant DNA, have occurred since the mid-1970s. These advances have given a new direction to established pharmaceutical, chemical, agricultural, and energy companies; of course, they have also given rise to the scores of new firms that are now such a distinctive characteristic of the biotechnology sector. Even in the earliest stages, it seemed clear that the dramatic progress in biotechnological research would have two important results: first, a wide assortment of new products with their own direct end-uses; and second, an even broader range of products with applications in the manufacturing processes themselves, technologies that would introduce revolutionary methods into the plant, the factory, the farm, and the refinery, as well as into the laboratory.

The current structure of the US biotech industry very much reflects its history. The first notable group includes companies such as Genentech and Cetus. These early pioneers have grown rapidly, established a firm market position, maintained their technological leadership, and are developing their own manufacturing facilities. A second dynamic segment involves even younger corporations built virtually on a pure technological base. They rely upon either patent royalties or research contracts obtained from larger competitors or outside financial sources. In some respects, this is both the most fragile and most innovative part of the industry. The third section of the biotechnology community consists of major firms with strengths either in the biochemical and medical underpinnings of the field or in areas of clear eventual application (such as energy or agriculture). Through contracts, marketing agreements, and joint ventures, these large companies have come to serve as a major source of capital for their smaller counterparts.

Until recently, biotechnology was an industry without commercial sales. Only one product for human use was on the US market before 1984. But the dramatic growth of investment in biotechnology clearly indicates the industry's direction — its 1982 level in the United States of $378 million jumped 64% to $619 million in 1983, with a similar increase to near $1 billion expected in 1984. Predictably, forecasts for an eventual biotechnology "market" range from the sublime to the ridiculous. Analysts expect anywhere from $15 billion to $100 billion in product sales by the year 2000, with incalculable additional benefits in terms of overall industrial and agricultural productivity. But most agree that a real surge in commercialized biotechnology will occur during the 1990s. The state of current research, the apparent rates of product development, and the normal lag between a product's introduction and its large-scale application all point to the next decade as a golden age for biotech. The following list indicates its broad future relevance by outlining some of the major end-uses where significant growth is expected:

1. Human Health
   a. Medicine / Disease Protection
      ● monoclonal antibodies
      ● drugs
      ● vaccines
   b. Nutrition and Diet
      ● food additives
      ● single-cell protein
2. Agriculture — Plant / Animal Production and Protection
3. Energy — Biomass, Oil Recovery
4. Environment — Soil Protection, Detoxification
5. Chemicals
   a. Organic Chemicals — biotechnological feedstocks and processes
   b. Production of Natural Metabolites
   c. Non-traditional Commodity Products (e.g., biopolymers).

Unfortunately, the atmosphere of excitement and opportunity that surrounds an industry with such promise has also spread to government agencies around the world responsible for their countries' domestic and international commercial policy. In all cases, these organizations have recognized biotechnology's considerable long-term potential from an economic perspective. Most have further concluded that a native capability in the field is essential to their long-term commercial growth and success. Then many reach the final and most problematic judgment that active government policies, in terms of tax treatment, antitrust guidelines, patent regulations, even outright subsidization of the private sector, constitute the most effective means of ensuring the rise of a strong biotechnology sector.

In the United States, government programs do play a significant role, but only in the basic research aspects of biotechnology. In 1983, the National Institutes of Health, by far the most active US government organization in this field, funded $571 million in biotech R&D, including some 4,500 grants in the fields of genetic manipulation, hybridomas, monoclonal antibodies, and immobilized enzymes. The research programs of other federal agencies (such as the Departments of Agriculture and Energy) will collectively provide an additional $120 million for work in biotechnology. Some of these monies are for in-house activity, but the majority are issued to universities and research institutes in the form of direct contracts.

Apart from these research-specific budget allocations, US government policies provide no direct support for biotechnology development. Of course, certain broader measures clearly hold benefits for industries of this type. The R&D tax credit offers relative advantages to any research-intensive field (although the measure is meaningless to most start-up firms since they report no taxable income). The more favorable tax treatment of long-term capital gains has also spurred the formation of new ventures, though again, the provision is impartial in its applicability. The Supreme Court has determined the applicability of US patent law to genetically engineered life forms. New antitrust guidelines recently passed by Congress for joint R&D legislate a more liberal environment for industrial collaboration in basic research; but once more, these criteria are indiscriminate, and they are backed by some of the most severe takeover restrictions in the world. All in all, US government philosophy and practice have dictated that biotechnology product development will occur in an unadulterated commercial setting.

As indicated, the general situation abroad is quite different. The immediate results are dramatic discrepancies between the policy environments at home and abroad for biotechnological development. Although competing together in a global market, each nation's industry will have developed under radically different conditions. And each such intervention will have skewed their composition and competitive capabilities.

There is a remarkable contrast between the make-up of foreign biotechnology sectors and their US counterparts. Aggressive start-ups and small innovative firms are virtually unknown outside of this country. While US tax changes in 1978 and 1981 certainly made such investments more attractive, they cannot fully account for the existence of these ventures or for their continued growth. But in both Japan and Europe, biotechnology is wholly dominated by, and in some cases the exclusive domain of, major established companies from related sectors.

Such a lopsided industry structure is only a predictable symptom of foreign targeting programs. In the case of Japan, "the Government has made the commercialization of biotechnology a national priority in order to ensure Japan's economic competitiveness and to decrease dependence on imports of foreign technology."* This commitment to foster domestic biotechnology has led to adoption of several key measures. Above all, the Japanese government has assumed a managerial role in industrial development and R&D. Its direct funding of basic research pales in comparison to recent levels witnessed in the United States, but this is primarily because its objectives are as much commercial as scientific. The very fact that the budget of the Ministry of International Trade and Industry (MITI) exceeds that of any other government agency, including the Ministries of Health and Education, testifies to this market orientation at even the most fundamental levels of policy planning.

MITI places a particular priority on the coordinated participation of the leading private companies in its major programs. The allocation of responsibilities along product or technology lines "reduces duplication" and allows firms to specialize in cultivating their own areas of expertise. This poses no competitive problems for the

---

*"Biotechnology," in *High Technology Industries: Profiles and Outlooks*, by Emily Arakaki, U.S. Department of Commerce, p. 94.

Japanese, since an industry's selection for promotion purposes generally entails removal of all meaningful antitrust constraints. Accordingly, fourteen top firms have joined forces in the Biotechnology Research Development Foundation (BRDF), through which they will administer and undertake extensive projects in three critical areas: recombinant DNA applications, bioreactor development, and mass cell culture technology. It is fully expected that such specialization will strengthen each company's eventual market position as well as its current technical capabilities.

In a similar targeting effort, the French government has selected biotechnology as the object of a special "mobilization program." Official objectives range from a 10 percent world market share by 1990 to successful completion of an ambitious three-year biotechnology research effort with joint government-industry funding of over $200 million. Further incentives, either in place or under current consideration, include special tax provisions, preferential or subsidized loans, and assistance in implementation of new research findings.

US concerns over the type of government involvement manifest in these and other examples are two-fold. First are the general distortions introduced and perpetuated by such extensive public participation in the development of particular industries. Free and open competition in the commercial marketplace bases its attractiveness on grounds of market efficiency and equity. At best, foreign targeting programs undermine both features; at worst they degenerate into a battle between nations instead of enterprises, between public policies instead of private initiatives.

Second are the costs of foreign targeting for a dynamic but immature biotechnology sector in the United States. While today's international market in biotechnology remains very limited, international flows of capital and technology have already assumed considerable importance in shaping the global competitive situation. We have already seen the effect of Japanese promotion efforts for their other domestic industry structures in such areas as computers, steel, and autos. The resulting business practices can have disruptive commercial consequences. Policies to limit (or, as the Japanese explain it, "rationalize") competition and the traditional large-firm orientation of the bureaucracies involved have resulted in a Japanese biotechnology sector with extensive resources. But, still, the primary limitation on Japanese development in biotechnology remains technical talent. What is the natural result, therefore, of this combination of inordinate financial strength and a severe lack of skills? A strategy of either direct buyouts or exchanges of capital for the technology of small innovative US companies is being pursued. These arrangements are currently in effect with 40 foreign firms, several of whom are Japanese.

The long-term implications of such patterns are difficult to assess. But there would not appear to be any extenuating circumstances to render our basic concerns invalid. Such fundamental government involvement can only disrupt the competitive balance. As commercialization proceeds, and the Japanese challenge in biotechnology intensifies, the significance of their acquiring these technological capabilities and cultivating these product/process specialties could become painfully obvious. In the meanwhile, one can only, through international pressures and negotiations, seek the dismantling of those targeting practices that have most blatantly furthered this effort—and hope for some success before it is too late.

22

Biotechnology Transfer— National Security Implications

John H. Birkner

Transfer of technology abroad can pose severe problems for the defense of the United States. Often the military importance of the technology is obvious, such as with weapons, explosives, or microprocessors. The Department of Defense (DoD), together with the Department of Commerce and the Department of State, has responsibility for monitoring critical technologies of potential use to foreign nations. Of greatest concern to these agencies, however, are those technologies whose military importance, at first glance, seems less obvious, such as biotechnology. The basis for adding biotechnology to the Militarily Critical Technologies List (MCTL) is the subject of this discussion.

This paper is divided into five parts: First, a few facts—those realities of the current technology export situation that people must learn to live with. Then issues—those concerns about which informed and reasonable persons could be expected to disagree. Next, some perceptions—the point of view that the Department of Defense could be predicted to adopt under various circumstances. Then a look ahead to a time when the United States will be faced, not just with hypothetical debate, but with hard decisions that could have a significant impact on the biotechnology business. Finally, some views of what government and industry, working together, should strive to accomplish.

The Facts

The first fact is that the laws involved reflect the government's obligation to monitor and, in some cases, block the sale abroad of militarily useful items and the transfer of technology which is imbedded in or otherwise accompanies exported equipment, materials, or information. If the export item has a direct connection with a military use, such as is the case with firearms or radar, the State Department has jurisdiction and could invoke the Munitions Control List to prohibit the export. This list has little bearing on biotechnology. Only if one happens to be in the business of manufacturing protective masks or the materials needed to manufacture toxic agents, and wishes to export such materials to a communist state or to one that sponsors terrorism would the Munitions List apply.

With regard to items other than weapons and other tools of war, the Department of Commerce administers the Export Administration Regulations, and has a catalogue of items—the Commodities Control List (CCL)—designating the kinds of things that require the Commerce Department's blessing in the form of a license before they can be exported; the rules vary according to the country of destination. It also falls to Commerce to deal with the export of technology that "could" be useful to a foreign nation for military application, as well as data about such technology that has not otherwise entered the public domain. This gray area of technology falls between what is obviously of military use, such as techniques for concealing an aircraft from radar, and those with no utility for war, such as the process for the enzymatic production of high-fructose corn syrup. These "gray area" technologies are often called "dual-use" technologies.

An instructive example of dual-use is the know-how involved in the manufacture of microprocessors, such as those found in everything from a simple laboratory balance or a pH meter to an air-to-surface missile or a submarine detection device. Current policy requires the licensing for export of not only manufactured microprocessor components, but also the equipment needed to produce them, and devices into which microprocessors have been incorporated. It also covers data, specifications, and training associated with producing them.

The mechanisms in place also have provisions for advice from the DoD about the wisdom of granting licenses. This is so, even though the technology may have only an indirect connection with the production of weapons or the prosecution of war. It is true that the laws involved here — the Export Administration Regulations — are in the process of being redrafted. The new version, however, is likely to differ only in the degree of control imposed and in the mechanisms provided for enforcement.

Those individuals involved with biotechnology are concerned in regard to the intentions of the government toward the export of bioreactors or high-capacity separatory devices. These examples are mentioned particularly because they incorporate dual-use technology that could be directly used in the manufacture of militarily useful materials. They can contribute not only to medical care, food production, and commodity chemical manufacture, but also can be used in military applications as specialized polymers, ultrasensitive detectors, or chemical and biological warfare agents.

Militarily Critical Biotechnologies

Some biotechnologies with dual uses have already been identified. As a result, some aspects of biotechnology are now considered to be militarily critical. The general definition of such technology is as follows: Military-critical technology is technology not possessed by the principal adversaries of the United States, or not available to them without restriction from outside the United States, that specifically contributes to the superior characteristics (performance, reliability, maintainability) of a military system, a significant component thereof, or a related strategic product. The idea is to identify and keep from these adversaries what they could use to improve their military capability.

There is a formalized and comprehensive list of such technologies that the DoD is required, by law, to create and make available to other government agencies. The jargon word for the DoD's list of dual-use technologies is the "MCTL," which is concerned with know-how and equipment that could be exploited, and describes such materials, processes, etc., in a manner that can assist in determining whether or not export licenses should be granted. This list of critical military technologies emphasizes arrays of design and manufacturing know-how; keystone manufacturing, inspection, and test equipment; and goods accompanied by sophisticated operations—application maintenance know-how. The list should be sufficiently specific to guide the determinations of any official exercising export licensing responsibilities.

One of the 18 categories in the MCTL is chemical technology. Five of the eight subsections for chemical technologies are relevant to biotechnology: polymeric materials, synthetic elastomeres, the manufacture and dissemination of toxic substances, detection and protective equipment, and manufacture and dissemination of biological and toxin materials.

Anyone who is involved with biotechnology and whose activities can result in the transfer of production know-how, manufacturing equipment, or technical data will be particularly interested in what the section on biological and toxin materials has to say. Equipment cited in the MCTL subsection includes production-scale high containment, biohazard decontamination, high-capacity bioreactors, high-capacity separation, solvent extraction/drying, microencapsulation, and spray nozzles with 1- to 10-microns pore size.

The assertion here is that if the United States supplies an adversary nation, such as the USSR, with technology related to these equipment items, then that nation's capability to produce and disseminate disease-causing warfare agents of biological origin will be enhanced. Unstated here, but inherent in the concept of MCTL, is the proviso that there is no point in blocking any technology the would-be recipient already has, or is likely to be supplied with, from some source other than the United States. The MCTL is meant to deal with that which is uniquely *the property of the United States* that should not be given away. It is not only the details of the MCTL with which biotechnologists should be concerned. Just as important is the question of when the MCTL should be invoked, and how it can best ensure that what it says about biotechnology remains pertinent as biological techniques and the uses to which they are put evolve over time. Opinions from industry in this matter will be valuable.

The Issues

So much for the facts. Biotechnology is stuck with export regulations, DoD has a role to play, and, because a broad spectrum of technologies is involved that potentially includes biotech, people will have to make the best of the situation. For purposes of illustrating the issues involved, consider a "what if" situation. It parallels an incident that happened in April 1984 shortly after it was documented by the United Nations that Iraq used both mustard and nerve agents in combat against Iranian troops. The United States, along with other nations, reacted by trying to block Iraq's access to the materials and equipment that it needed to produce chemical warfare agents in quantity. Much of the embargoed material was dual-use in character in that it also has utility in the manufacture of insecticides and other chemicals. What if it were to become apparent that a foreign power was engaged in developing, producing, stockpiling, or using chemical and biological warfare (CBW) agents derived from biotechnology techniques or equipment items provided to that power by the United States as a result of an export sale?

If such a thing happened, how should it be regarded? As an inevitable outcome that must be accepted, like death and taxes? As a failure on the part of those responsible for policing such activity as dictated by present laws? As an indication that the country must redouble its efforts to create new policies that will stem the flow of technology by further restricting exports? Remember, the mechanism is there to react, and to use, if one chooses. It has happened with truck production equipment, with bearing technology, with solid-state production know-how, with computers, and with radar. The list of examples is long; the expectation is that it will grow longer and could eventually include any technology which the United States has that contains innovative qualities that others would like to obtain by borrowing, buying, or stealing.

What should be done if commonly available US-derived technology was found to be the basis for the development of a new foreign military threat? Nothing? There are many who would wish otherwise. Where industry and government disagree is on the practicality of comprehensive export restrictions. The skeptics offer the following: "If we don't sell it to them, someone else probably will." "No harm in selling it; they already have similar technology." "Not selling it would only slow them down; it won't stop them." "It's okay to give them this; it's not our latest generation. We will keep one step ahead." And, finally, "biotechnology has nothing to do with war."

Fine. These statements can be relevant for developing export policies, provided they are authentic reasons, and not just convenient excuses. It's really no wonder that excuses for limiting export controls are legion. After all, the impacts are immediate in business — one is talking about real profits, right here, right now. The prospect of harm to national security is a postulated danger occurring somewhere else at some time in the future.

With technology transfer in the electronics area, however, it turns out that the danger is not just postulated; it is currently manifested right off American shores, and it is causing trouble now in the form of Soviet anti-submarine warfare detection devices. How ought people to respond if biotechnology transfer brings on some sort of parallel development? This argument from analogy is invoked by those who favor more restriction. They say foreign governments have shopping lists for technology, elaborate networks of operatives, dummy corporations for disguising the ultimate destinations of exported goods, and even agents with instructions to steal if all other means fail. In the face of all this, those concerned about technology loss believe that both naive and unscrupulous exporters alike can cause great harm unless there is intervention by the government.

Perceptions in DoD

It is also important to consider the particular perceptions of those in the Department of Defense. The military can be expected to want restraints on exported items that could enhance the CBW offensive capabilities of our potential adversaries, their self-protection capabilities against their own weapons, and technology that could help them produce new and unique materials or develop other types of exotic applications that have been postulated. DoD, moreover, wants to ensure that whatever benefits can be derived are applied to bolster US capabilities. DoD wants it invented here, and sold here, and the benefits to be realized here. What DoD does *not* want is for technology to go to potential enemies, most especially if the US military does not even know about the applications. Further, the United States would expect its allies to share these concerns, and to cooperate with these efforts. From DoD's perspective, restraining the loss of militarily sensitive technology would probably never be as effective as one would wish. But, at the very least, DoD would like to keep close track of which technologies do interest any adversaries and what they are receiving.

A Look Ahead

So where do biotechnology transfer and national security concerns ultimately lead? It is now prediction time. First, in general terms, the intelligence community is looking hard for evidence about what the very large, very diverse, and very secretive research and development establishment in the USSR is intent on achieving in the way of a military technological surprise that would work against the interests of the United States. This applies to technology of all kinds, including biotechnology. The advent of genetic engineering, advances in protein biochemistry, powerful immunological methods, improved bulk purification techniques, and ultrasensitive chemical identification procedures have ushered in a new era in basic biology discovery and biotechnology applications. Some of these are in the military arena. In the area of chemical warfare defense, American R&D is already underway to employ gene-splicing techniques to devise new forms of antidotes for nerve-agent poisoning.

The United States should not be complacent about potential adversaries and their capability for applying biotechnology. This is particularly true for the USSR. What the Soviets accomplish is driven as much by their priorities as it is by their abilities. The biomedical field is a good example. The United States should be careful in drawing unwarranted inferences from appearances. Just because the Soviet approach to civilian-sector health care delivery seems primitive, the United States should not conclude that their technical establishment is unable to apply new biomedical technology. Instead, it is a matter of what the Soviets *choose* to afford, because other sectors, such as defense, are preferentially supported.

How much and what kinds of military applications do foreign military planners assess biotechnology to have? In the case of the Soviets, have they attempted innovative R&D approaches? Are they concerned that the West may eventually blunt the effectiveness of their current chemical arsenal? There are also biological and toxin warfare possibilities to consider. The Soviet Union is assessed to have an active R&D program to investigate and evaluate the utility of biological weapons. This effort violates the Biological and Toxin Weapons Convention of 1972, which was ratified by the USSR. The Convention bans the research, development, production, and possession of biological agents and toxins for warfare purposes.

There are at least seven biological warfare centers in the USSR that are under strict military control. There is an apparent effort on the part of the Soviets to transfer selected aspects of genetic engineering research to their biological warfare centers. For such purposes, recombinant DNA techniques could open a large number of possibilities. Normally harmless, nondisease-producing organisms could be modified to become highly toxic and to produce effects for which an opponent has no known treatment. Other agents, now considered too unstable for storage or biological warfare applications, could be changed sufficiently to become effective.

Of greatest concern is that biotechnology R&D applied to such purposes would be very easy to conceal. Unlike high-energy physics experiments or the construction and testing of weapons delivery vehicles, new biotechnology research efforts devoted to military objectives would tend not to reveal themselves. Facilities, equipment and personnel devoted ostensibly to food or drug production could easily be turned to military biotechnology R&D tasks.

There is a better-than-even chance that the United States will find out that the USSR is actually pursuing offensive military applications for biotechnology that dwarf the defense-oriented programs being pursued by Western nations. In such a circumstance, the United States would probably elect to take action to halt technology loss. The United States is not obliged to wait until confronted with some instance of actual battlefield use.

The actions that might be entered into could take several forms that would have an impact on the export of equipment, production know-how, and technological data. Opinion could become strongly polarized in the process. On one side would sit the manufacturers and exporters with goods to sell, and considerable contempt for the government's case. On the other side would be the most zealous of those from DoD, and perhaps Commerce and Customs, with the task of rooting out all those who would dare to cooperate with the enemy. In the middle would be the less vocal, more temperate individuals trying to be fair and reasonable. Eventually in such an encounter the government would probably prevail as it has in the microprocessor situation.

What Can Be Done

Whatever the future may hold, one thing is certain. The less industry–government interchange that takes place now, the more acrimony there will be if these predictions come true. Now is the time that everyone should work towards making the MCTL right, in anticipation of the day when it will have to be used in earnest. It is one of the prudent contributions that can be made to help protect future profits. In addition, it is important to find out how technology may be turned against the United States. That is a prudent contribution that can be made to protect the future security of this nation.

There are many mechanisms and ongoing forums in which cooperation can occur. Specifically, DoD would like counsel from the industrial and academic communities on such topics as

- How best to assess the implications that biotechnology has for national defense;
- The degree of foreign interest in those applications that have military utility;

- What to include in the MCTL;
- How much transfer is now occurring;
- How to improve DoD access to militarily useful applications; and
- How DoD can help to ensure that the United States remains in a position of innovative biotechnology leadership.

Suppose a scientist from the United States attempted to get a Soviet Communist Party official or any person with influence and/or authority to tell him about what the Soviet military is doing with biotechnology. Would a straight answer be forthcoming? Not likely, and everyone knows why. To a greater degree than almost anywhere else in the world, deception and lack of forthrightness is a way of life in the government bureaucracy of the USSR. On the other hand, the prospects for discovering what the United States is doing in the area of military applications of biotechnology are very good because it is public knowledge and the responsible officials in the agencies involved are committed to a policy of open disclosure. Consider this difference between the two nations and their respective governments. Then consider with whom one should be cooperating, and the care one should take in dealing with foreign customers.

23

Biotechnology — Export Controls

Hank Mitman

Today the Office of Export Administration faces a critical issue. To what extent is it in the best interests of the United States to regulate the export of biotechnology? The question is not easily answered for two reasons. First, the technology is so new that it is difficult to evaluate. Second, export regulations must not affect the basic research underlying biotechnology. To deal with these problems, the Office of Export Administration has asked industry representatives and members of other agencies to review current export controls.

The primary function of the Office of Export Administration (OEA) is to implement the Export Administration Act of 1979. This act expired on September 30, 1983; it has, however, been continued in effect under the International Emergencies Economics Powers Act. (Congress reauthorized the Export Administration Act in 1985.)

The functions of the act are three-fold:

- It controls exports "which would make a significant contribution to the military potential of any country which would prove detrimental to the national security of the United States."
- It controls exports, where necessary, which "further significantly the foreign policy of the United States or fulfill its declared international obligation."
- It controls exports "where [it] is necessary to protect the domestic economy from excessive drain of scarce materials and to reduce the serious inflationary impact of foreign demand."

The various commodities controlled by OEA fall into ten groups, with those currently pertaining to biotechnology falling within groups 7 and 9. Group 7, for instance, includes products of certain fermentation processes, such as acids and alcohols, some of which require a license to reach specified destinations. Group 9 entries include microorganisms; viruses of human, veterinary and plant origin, as well as bacteria, fungi and protozoa; these entries require validated licenses to all destinations. Also, Interpretation 28 in Part 399.2 of the regulations controls certain identified microorganisms and protozoa that require a validated license for shipment to Libya, North Korea, Vietnam, Kampuchea, and Cuba, but which may be exported under general license to all other destinations. Based on this somewhat complex approach applicable to commodities of interest, a review and update of these controls would appear to be warranted.

According to a recent survey by Sittig and Noyes in *Genetic Engineering and Biotechnology Firms Worldwide, 1983/84*, there are approximately 1,200 biotechnology firms worldwide. Their rapid growth in the application of recent discoveries in biology to commercial products indicates that this industry may have a multi-billion dollar commercial impact before the end of this century. In this context, it appears that the best means of maintaining US leadership in this emerging industry, according to several government studies, would be to develop coherent government policies.

Recommendations

Among the recommendations made by the White House Office of Science and Technology Policy (OSTP) were that the Department of Commerce bolster its expertise for the purpose of updating and streamlining its controls in the biotechnology area and that the department establish close coordination among the Departments of Defense and State and the intelligence community. It was also recommended that, if sufficient interest became evident in the biotechnology industry, a Technical Advisory Committee (TAC) should be established to address areas of specific concern to this industry and to include comments on the Department of Defense Militarily Critical Technologies List (MCTL).

Based on the OSTP findings developed in 1983, the Department of Commerce has begun an implementation program. For example, the National Institutes of Health has detailed one of its senior research scientists to the Department of Commerce to assist the department in establishing a TAC and to initiate a review of its current export controls. It is the intent of the department that this TAC be composed of representatives from both small and large US biotechnology, pharmaceutical, and scientific equipment companies.

Recognizing that the federal government regulatory procedures directly affect the biotechnology commercial sector, the Department of Commerce in 1984 held a high-level conference chaired by Lionel Olmer, undersecretary for International Trade Administration. The object of that meeting was to establish a dialogue between senior industrial executives and officials of government. Five major subject areas were presented and discussed at that meeting. These included:

- *Patents*, including (a) enhancing the quality and quantity of Patent Office examiners in areas related to biotechnology; (b) strengthening the Plant Variety Protection Act; (c) supporting legislation to restore patent life lost during the government regulatory approval process; (d) working toward uniform worldwide practices in patent issuances in biotechnology; and (e) developing worldwide reciprocity in patent rights.
- *Statutory changes in US drug export policy*, including a potential change in the US Drug Export Policy regulations to permit the export of new drugs intended for human consumption, under the appropriate conditions, prior to receiving approval for human use in the United States.
- *Economic development*, including the development of technology transfer mechanisms to identify promising basic research findings suitable for applied work and development of a mechanism to provide funding for applied research.

- *Public safety*, including the resolution of the government safety oversight role of commercial recombinant technology.
- *Export controls of biotechnology*, including the development of export control capabilities in biotechnology by establishing a Department of Commerce Technical Advisory Committee which will function to: (a) enhance industrial and public awareness of the Export Administration Act as it pertains to biotechnology; (b) streamline and update export regulations affecting biotechnology with close cooperation between the trade development and export control elements of the Department; (c) enhance the department's interaction with other federal agencies which may have an impact on biotechnology trade, such as the Food and Drug Administration, Environmental Protection Agency, Department of Defense, National Institutes of Health, Department of Agriculture, and Department of State; and (d) keep the Department of Commerce abreast of the latest developments in biotechnology.

The OEA Agenda

The Office of Export Administration is drawing up an agenda that will be presented to the Biotechnology Technical Advisory Committee as soon as it is officially constituted. (The committee held its first meeting in April 1985.) Topics to be covered will be a review of current export regulations and development of a concerted public relations approach so that all parties will be fully aware of the existence of export regulations.

Specific areas for review in reference to possible export controls will be products of biotechnology processes, technology transfer, biosensors, and biochips. Additionally, other means of safeguarding national security and promoting trade development will be explored, and perhaps by these mechanisms retain the current US technological lead.

Until recently, the main thrust of the department's effort in export control areas has been to prevent the flow of US technology to the Soviet and Eastern Bloc countries. Currently, other options are being considered in the implementation of the overall policy, such as controlling all transfer of biotechnology. Four considerations in this effort include:

- Dual Use — does this technology have both military and civilian applications?
- Military Criticality — how critical is the technology for improving the recipient's military capability or strategic posture?
- Foreign Availability — if the technology is generally available from foreign sources, can the US achieve any strategic goals by regulating its transfer?
- Effectiveness Transfer — which technology transfer control mechanisms are, in fact, feasible?

Each of the above issues needs to be explored in more detail to get a better understanding of its implications.

The Office of Export Administration does not control fundamental research or basic science. It is neither OEA's intent nor desire to control activity in this area, nor would it be in the best interest of the United States or science to do so. In other words, the Department of Commerce's regulatory process specifically excludes basic research.

It should be recognized, however, that the transition point from basic to applied research needs to be addressed in relation to its impact on national security. Consequently, if the product of the applied research has dual use (commercial and military), then, in such instances, controls may be required, depending on criticality, impact on military balance, foreign availability, and other factors.

Criticality and Capability

Once a decision is reached that a technology is of dual use, the next question to be considered is how critical is it militarily? Does the technology have the potential to change the military balance or to provide a new capability to the recipient country? Answers to these questions should aid in interpreting the Militarily Critical Technologies List.

If the technology meets the criteria of dual application and criticality, its foreign availability should be evaluated. Foreign availability means the availability to the proscribed countries of commodities subject to US export control. Such commodities may be available from any source, but must be of comparable and sufficient quantities to be meaningful. The objective of the Foreign Availability provision of the act was to minimize the adverse effects of the export control system on US competitiveness.

Having made a determination in respect to the first three tests — dual use, military criticality, and non-availability outside the United States — some thought should then be given to how effective the controls would be if implemented. If the controls are to be effective, their implementation should be through selection of a mechanism from several possible modalities that offer the best potential for effectiveness. This would clearly need careful consideration.

This discussion provides a status report of OEA's current situation and future direction in the field of biotechnology. Input from the biotechnology community is essential. There is probably general agreement that national concerns should be addressed whenever new fields of expertise emerge, and that everyone desires to achieve equitable federal policies in the best interest of the United States. These are the common objectives that are now in the process of being formulated.

24
National Policy and Biotechnology in the United States

Irving S. Johnson

Unlike other countries, the United States appears to be floundering around for a national policy on biotechnology, while squandering its fragile competitive lead in this area. In comparison, most other developed countries have targeted biotechnology as a national goal. These countries include the United Kingdom, France, West Germany, the Soviet Union, and—even more vigorously—Japan. The national efforts of these countries have ranged from modification of guidelines for carrying out biotechnology research to legislative assistance, financial support of private companies, and dismantling of unneeded antitrust legislation.

That US policy is not better organized is surprising. With the exception of hybridomas, many of the original discoveries leading to biotechnology have been made in the United States. Furthermore, the United States has had a long history of funding basic research. The United States has also shown flexibility in the past in developing the recombinant DNA Guidelines.

But comparing the efforts of the United States with those of the United Kingdom underscores the differences. In the United Kingdom, the recombinant regulatory authority includes representatives from industry. This authority has recently been merged into the Department of Trade and Industry that also contains a biotechnology unit. This unit seeks advice on the utilization of natural resources while supporting the efforts of private companies. These companies range from small ones such as Celltech, to giants such as ICI. More recently, this combined authority has moved to organize agricultural biotechnology in the United Kingdom.

The situation is similar to Japan, where biotechnology has been targeted as a national goal. As a result, the government has coordinated the efforts of industry, academia, and the government. Japan has formed a chemical industry consortium for biotechnology research with fourteen separate companies participating. The Japanese government has both encouraged and assisted interest and investment by Japanese firms in transfer of this technology from biotechnology firms in the United States. Japan has developed programs to strengthen bioprocessing and fermentation, in which they are already proficient. The Japanese have frankly admitted that whereas they may be five years behind in biotechnology, they intend to make the difference up quickly by scrapping older technologies and improving on new technology imported from the United States as part of their first five-year plan.

There is a similar theme in the approaches of both the United Kingdom and Japan as they develop a national science policy for biotechnology: industry as a partner, not just a protagonist.

The US Approach

Contrast this approach to recent developments in the United States. Federal agencies are divided over who has responsibility for biotechnology. The National Institutes of Health (NIH), which originally established guidelines for biotechnology research, is now under pressure to give up that authority to the Environmental Protection Agency (EPA). EPA has not been given any specific authority in this area, but has loosely interpreted the federal Insecticide, Fungicide, and Rodenticide Act to encompass DNA as a potentially hazardous chemical.

The Department of Defense, the Department of Commerce, and others debate the extent of restrictions on exporting biotechnology to other countries. American courts are filled with lawsuits challenging a host of biotechnology issues, ranging from deliberate release of microorganisms to patent questions, particularly those involving universities and industry, as well as questions on the rights of patients to share in uses of biotechnology created from their body tissues. In all of these areas, industrial participation in formulating national policy is restricted, if not nonexistent.

The Obvious Steps

What can be done about such problems? Even within the context of national science policy in the United States, some steps are becoming increasingly obvious:

● Industry must be given a more representative voice in contributing to debates on all facets of biotechnology policy.
● Federal assistance should be targeted to bioprocessing and applied microbiology centers, possibly by funding through universities.
● Federal support of basic research should be increased, particularly in agriculture where the United States may already be lagging behind. Some US fellowships used to train foreign scientists at leading biotechnology centers in this country should be converted. Instead, the funds could be used to train American scientists at foreign technology centers.
● The oversight by the NIH's Recombinant Advisory Committee should, alone, be continued. To do otherwise would simply favor foreign competition.
● EPA should not be permitted to regulate biotechnology at any stage, except at the point of production. To do otherwise would basically sabotage research efforts and seriously interfere with freedom of scientific inquiry.
● Intellectual property law, such as protection of plant varieties and parts as intellectual property, should be strengthened by the formation of a scientific advisory committee with sufficient industry representation.
● Regulations on transfer of biotechnology should be kept to a minimum. Where exportation does occur, adequate returns should be anticipated.
● Congress needs to reexamine antitrust legislation in the biotechnology field.
● The tax code should be clarified and updated to give R&D incentives in the biotechnology area.

The need for reassessing and developing a national biotechnology policy is obvious. The usefulness of this technology for the United States may otherwise become overshadowed by the potential use of biotechnology for germ warfare. Reactions to such usage have already led to calls to regulate the transfer of the technology to other countries in ways that could be harmful to further domestic development of biotechnology. For example, petrie dishes, centrifuges, ion-exchange columns, and fermenters are widely available outside the United States, and placing export restrictions on them would have a detrimental effect on the ability of US companies to compete internationally.

In addition, application of biotechnology for purposes of biological warfare can only be controlled by *verifiable* international agreement. It should be noted in this context that it is difficult to conceive of the use of biotechnology to produce an agent more deadly or toxic than the original pathogenic organism found in the environment. Furthermore, one must remember that regulations cannot prevent either immoral or improper behavior. But a logically developed, multifaceted approach to a national policy will keep the United States in the lead of biotechnology development, where it belongs, for years to come.

25

International Biotechnology Transfer

Nanette Newell

Transfer of technology is trade dependent with this country's access to world markets and international competitiveness. This dependency on trade is particularly true for selling pharmaceuticals abroad — the initial products of commercialized biotechnology. The interdependence of trade and technology transfer must be considered, therefore, when export controls are evaluated. Thus, export controls for national security reasons are likely to affect both trade and technology transfer, rather than just technology transfer alone.

Increasingly, the success of companies depends on their ability to gain access to world markets. This is especially the case for the pharmaceutical industry, which is truly international. Wholly owned subsidiaries, foreign manufacturing plants, and cross-licensed patents are more the norm than the exception in the pharmaceutical industry. Thus, if a new biotechnology company in the United States wants to develop pharmaceuticals, it must acknowledge (and participate in) these international arrangements in order to compete effectively in world markets.

With the development of a new technology, there are four ways to gain access to world markets:

- License technology to a foreign company;
- Form a joint venture with a foreign company;
- Manufacture a product through a wholly owned subsidiary in a foreign country; or
- Export a product.

These methods of international market penetration are listed in the order of the most technology transferred to the least.

Licensing

Most new biotechnology companies, in their early stages of development, choose licensing as the preferred means of market access. There are many reasons for this choice. Licensing may be necessary if a small company does not have manufacturing and marketing expertise. This situation is generally the case for small, new companies because they have limited capital. If licensing is selected as the preferred method of getting one's product to the market, then the company chooses between licensing to a large domestic firm or a foreign firm.

Since most US biotechnology companies want to preserve the large US market for themselves, many choose to license to foreign firms where the arrangements usually give marketing rights for many foreign countries to the foreign firm in return for royalties on sales. The small firm retains marketing rights in the United States and possibly some foreign countries. These arrangements usually confer long-term benefits on the foreign firm. It should be noted that foreign firms were much more interested in licensing arrangements than were US firms when these new biotechnology companies were first launched. Since these new companies needed money, there was certain pressure to make arrangements with foreign firms.

Joint Ventures

As young biotechnology companies gain more secure financial and scientific footing, joint ventures become a preferred method of gaining access to world markets. Less technology is transferred to the foreign partner than in a licensing agreement, and the US partner is still involved in corporate decision-making. Joint ventures may also be used when a foreign country does not allow wholly owned subsidiaries.

Subsidiaries

Wholly owned subsidiaries are used extensively by the pharmaceutical industry because they give good access to foreign markets. Technology is transferred when employees move from one company to another.

Export

Exporting the product directly transfers the least amount of technology. There can be, however, many tariff and nontariff barriers to trade that discourage this option. Additionally, domestic regulations may interfere with the export of a product. For instance, regulations in the United States prevent the export of drugs that have not first been approved for use in the United States.

In biotechnology, as with other technologies, technology transferred in or out of a country is difficult, if not impossible, to qualify or quantify. Most biotechnology analysts, however, would agree that up to now, the United States has probably transferred more biotechnology out of the country than it has imported. The US export of technology, however, should not be overemphasized. The US market is large, and foreign firms usually consider it essential to gain access to this market. Thus, as biotechnology products reach the market, foreign firms will begin transferring technology into the United States by the same mechanisms mentioned earlier.

This discussion leads to another point. Often, a parochial attitude is taken that all good technology originates in the United States. In the field of biotechnology, this is definitely not true. The scientific community is truly international, and many important contributions to biotechnology have been made by foreign scientists. For instance, the discovery of monoclonal antibodies by Argentinean Cesar Millstein in England (for which he received the 1984 Nobel Prize in Medicine) is only one of numerous examples of major discoveries made within the international scientific community.

In addition to many foreign contributions in basic science areas, Japan is often recognized as the leader in bioprocessing, and Sweden and other European countries produce some of the best separation equipment used in biological manufacturing. If the United States were to control high technology exports in this field, it would undoubtedly harm the competitive position of US companies because of likely retaliation by foreign countries. By maintaining strong bilateral relationships, US companies will realize significant benefits.

Because trade and technology transfer are intertwined, it is well to be aware of congressional activities concerning the Export Administration Act (EAA), which expired in the fall of 1983. A new bill is being considered by Congress. The Reagan administration favors more stringent controls, generally, on the transfer of high technology, and the Senate version of the EAA reflects this belief. (Congress reauthorized the act in 1985.)

Additionally, a critical change in thinking about export controls has occurred in the administration. There has been a shift from the regulation of the products themselves to the regulation of the technology that contributes to those products. There are questions about how and whether one can actually regulate technology transfer in this manner, but there is evidence that the regulation of attendance and papers presented at scientific meetings could be one method. Besides raising First Amendment concerns, this trend could substantially slow the progress of scientific research by preventing the exchange of data and ideas within the scientific community.

Conclusion

In conclusion, it must be recognized that scientific developments in biotechnology will be fostered by open communication channels in the worldwide scientific community, and any effort to close those channels can do nothing but harm the competitive position of US companies. Further, in any high technology business, technology transfer and access to world markets go hand in hand. Any inhibition of the ability of US companies to gain access to world markets will decrease US competitiveness.

About the Editor

Dr. Perpich, Vice President for Planning and Development, Meloy Laboratories, Inc., is a psychiatrist and attorney. He graduated from the University of Minnesota Medical School in 1966, completed his internship at the University of Minnesota Hospitals and then his residency in psychiatry at the Massachusetts General Hospital and the National Institute of Mental Health. A mental health career development award from the National Institute of Mental Health allowed Dr. Perpich to combine his psychiatric and legal training with a year as a Congressional Fellow on the Senate Committee on Labor and Public Welfare's Subcommittee on Health, and a year as a law clerk to David Bazelon, then Chief Judge for the US Court of Appeals for the District of Columbia Circuit. Dr. Perpich received his law degree from the Georgetown University Law Center in 1974.

From 1974 to 1976, Dr. Perpich served at the Institute of Medicine, the National Academy of Sciences, where he was responsible for developing a program in legal and medical ethics. From 1976 to 1981, he held the post of Associate Director for Program Planning and Evaluation at the National Institutes of Health under NIH Director Donald S. Fredrickson. Dr. Perpich's first assignment in that position was to direct the staff effort on developing the NIH Recombinant DNA Research Guidelines and related Federal legislative, judicial and executive policies. Other responsibilities included the development of planning processes and Federal research strategies.

During his last two years at the NIH, Dr. Perpich worked to bring industry further into the government-university partnership in health research — especially in biotechnology. He then was Vice President for Corporate Planning and Government Affairs at Genex Corporation (1982–1983).

Dr. Perpich is a fellow of the American Psychiatric Association and a member of the Bar of the District of Columbia.

About the Contributors

David L. Bazelon is Senior Circuit Judge of the United States Court of Appeals for the District of Columbia (retired). He is a graduate of Northwestern University Law School.

John H. Birkner is presently assigned to the Allied Forces South NATO staff in Italy. Dr. Birkner, a Colonel in the US Air Force, received his doctorate degree in environmental biology from Colorado State University.

Ronald E. Cape is Chairman and Chief Executive Officer of Cetus Corporation. He holds an A.B. in Chemistry from Princeton University, an M.B.A. from Harvard University's Graduate School of Business Administration, and a Ph.D. in Biochemistry from McGill University.

Alexander Morgan Capron is Topping Professor of Law, Medicine and Public Policy at the University of Southern California. Professor Capron is a graduate of Swarthmore College and Yale Law School.

J. Leslie Glick is Chairman of the Board and Chief Executive Officer of Genex Corporation. He received his undergraduate and graduate degrees from Columbia University. Dr. Glick also serves as President of the Industrial Biotechnology Association.

Harold P. Green is Professor of Law and Associate Dean for post-J.D. Studies at the George Washington University School of Law. Mr. Green holds an undergraduate degree in economics from the University of Chicago and is a graduate of their law school.

Peter Barton Hutt is a partner in the Washington, D.C. law firm of Covington and Burling. Mr. Hutt received his undergraduate degree from Yale University, his law degree from Harvard University, and his master's in law from New York University.

Irving S. Johnson is Vice President of the Lilly Research Laboratories, a division of Eli Lilly and Company. He earned his undergraduate degree and his doctorate degree in experimental biology from the University of Kansas.

Donald Kennedy is President of Stanford University. He received his undergraduate, master's and doctorate degrees from Harvard University. His academic career has been largely at Stanford, where he was Professor and Chairman of the Department of Biological Sciences.

Richard M. Krause is Dean of the Emory University School of Medicine and Woodruff Professor of Medicine. He was formerly Director of the National Institute of Allergy and Infectious Diseases (NIAID) at the National Institutes of Health. A physician, immunologist and microbiologist, Dr. Krause received his training at the Case Western Reserve University Medical School.

Hank Mitman is Director of the Capital Goods and Production Materials Division of the US Office of Export Administration. He is a graduate of Gettysburg College and did graduate work in physics at the Johns Hopkins University.

Nanette Newell is an industrial biotechnology consultant in San Francisco and former Director of Research Administration at Calgene, Inc. Dr. Newell received her undergraduate degree in chemistry from Lewis and Clark College, and her doctorate degree in biochemistry, and cellular and molecular biology from the Johns Hopkins University School of Medicine.

David W. Plant is a partner in Fish and Neave, a New York patent law firm, and chairs the Committee on Patents of the Association of the Bar of the city of New York. He received his undergraduate and law degree from Cornell University.

Clyde V. Prestowitz, Jr. is Counselor to the Secretary of Commerce for Japan. Mr. Prestowitz is a graduate of Swarthmore College and holds a master's degree from the University of Hawaii's East-West Center and Keio University in Tokyo.

Harrison Schmitt is a former senator, former astronaut and key formulator of science policy. He holds an undergraduate degree from the California Institute of Technology and a doctorate in geology from Harvard University.

Lewis Thomas is Chancellor at the Memorial Sloan-Kettering Cancer Center. Dr. Thomas is a graduate of Princeton University and Harvard Medical School.

Nicholas Wade is a member of the editorial board of *The New York Times*. Mr. Wade was formerly the deputy editor of *Nature* and a member of the news staff of *Science*.

Sir Gordon Wolstenholme is an Honorary Fellow of the Royal Society of Medicine and a Foreign Honorary Member of the American Academy of Arts and Sciences. Sir Gordon served as chairman of the Genetic Manipulation Advisory Group, President of the Royal Society of Medicine, and member of the General Medical Council.